# Hubble

## The Mirror on the Universe

# A FIREFLY BOOK

Published by Firefly Books Ltd. 2011

First printing

Publisher Cataloging-in-Publication Data (U.S.)
Kerrod, Robin.
    Hubble : the mirror on the universe / Robin Kerrod and Carole Stott.
3rd ed.
[224] p. : col. photos. ;  cm.
Includes index.
Summary: Latest images from the Hubble Space Telescope, accompanied with text
explaining their astronomical significance, details on the Telescope and a timeline of
landmarks in astronomy.
ISBN-13: 978-1-55407-316-0  (pbk.)
1. Hubble Space Telescope (Spacecraft).  2. Orbiting astronomical observatories.
3. Outer space — Exploration. I. Stott, Carole.  II. Title.
520.22/2 dc22    QB500.268.K477    2011

Library and Archives Canada Cataloguing in Publication
Kerrod, Robin
            Hubble : the mirror on the universe / Robin Kerrod & Carole
Stott. — 3rd ed.
Includes index.
ISBN 978-1-55407-972-8
            1. Hubble Space Telescope (Spacecraft).  2. Space astronomy.
I. Stott, Carole  II. Title.
QB500.268.K47 2011    522'.2919    C2011-901251-0

Published in the United States by
Firefly Books (U.S.) Inc.
P.O. Box 1338, Ellicott Station
Buffalo, New York 14205

Published in Canada by
Firefly Books Ltd.
66 Leek Crescent
Richmond Hill, Ontario L4B 1H1

This book was conceived, designed, and produced by
Quintet Publishing Limited
6 Blundell Street
London N7 9 BH

Editor: Anya Hayes
Designers: Rod Teasdale
Managing Editor: Donna Gregory
Art Director: Michael Charles
Publisher: Mark Searle

Color separation  in Singapore by Pica Digital Pte Ltd

Cover images courtesy of NASA and STSci.

Printed in China

OPPOSITE
**Return to orbit
The Hubble Space
Telescope imaged
from the space shuttle
Atlantis as the two
spacecraft separate in
May 2009. They had
been linked together
during the final
mission to service
the telescope.**

# Hubble

## The Mirror on the Universe

# Contents

# Foreword

Everyone who works on the Hubble Space Telescope program encounters people in his or her everyday life—the dentist, the auto mechanic, a stranger sitting next to you on an airplane—who asks us what we do for a living. When we say, "I work on the Hubble Space Telescope", invariably their eyes light up and they say, with a tone of excitement, something like, "Wow, that's so cool," or more recently, "Gee, I hope you can save Hubble." I'm often asked what it is about Hubble that causes it to be so enthusiastically embraced by so many people—not just scientists, but people from all walks of life. Many explanations have been suggested, probably all true to some extent.

Here's my take on it: we are privileged to be the first generation of *Homo sapiens* to gain a clear and deep view of the visible Universe. And what we see "out there" is staggering in its beauty, awesome in its scale, and shocking in the way it has upended our preconceived notions about how nature works. You don't have to be a scientist to grasp this. Any thinking person who has come in contact with Hubble images and Hubble discoveries seems to find exhilaration in the notion that our place in the grand scheme of things is now better defined than in all of human history to date.

There are some other important factors. The Hubble Space Telescope belongs to all of humanity. It is an international facility that any scientist in any country can competitively propose to use. In the process of peer review, the best ideas win time on the telescope, regardless of where they came from. At the same time, any person on earth who has access to the internet can peruse the entire archive of Hubble data, not to mention the large store of Hubble imagery and other material specifically aimed at a non-professional audience. Any student at any level at any school in the world at any time can gain knowledge and inspiration from Hubble observations. For the many of us who are lifelong fans of science fiction, Hubble is the surrogate starship that transports us across the Universe when there is, as yet, no other way to make the journey. It gives flight to our imagination and creativity.

Finally, we as human beings can take justified pride in the fact that we have created and used Hubble for entirely peaceful purposes in a world that suffers continuous conflict and pain. Hubble is noble. And we made it!

David S. Leckrone
Senior Project Scientist
Hubble Space Telescope
NASA, Goddard Space Flight Center
June 5, 2007

PAGE 4
**Galactic Collision**
**Two spiral galaxies, known together as Arp 272, are linked by their curved arms. The colliding galaxies are about 450 million light-years away and part of the Hercules Cluster of more than 100 galaxies. The cluster is, in turn, part of the Great Wall, one of the largest known structures in the Universe.**

LEFT
**Final mission**
**Space Shuttle Atlantis' payload bay against the backdrop of black space and the thin line of Earth's atmosphere.**

# Hubble: The Story So Far

For more than two decades the Hubble Space Telescope (HST) has enabled us to see and understand our Universe more clearly. It has also helped us to place our home and our lives in a wider context, balancing our concept of the vastness and permanence of planet Earth with the realization that we actually live on a rather small planet and that half the planets in our solar system are much bigger. Hubble has allowed us to probe the childhood of our Universe and witness the formation of galaxies, investigate many of the billions of stars in our own galaxy, see the extraordinary spectacle of star birth and death, and explore objects in our solar system.

The HST Flight Operations Team is located at the Goddard Space Flight Center in Greenbelt, Maryland. The telescope is observing 24 hours a day, 7 days a week, the only respite being when the instrument is visited by astronauts during the servicing missions. The amount of work done by the Hubble Space Telescope is prodigious: during its first twenty years of operation (May 1990–April 2010) it has taken more than 570,000 astronomical images of 30,000 celestial bodies. The exposure times of each image varies drastically depending on the faintness of the body, but this production rate is equivalent to an average of 3.4 images per hour. The astronomer whose observation program is being carried out by the telescope has exclusive rights to analyze the image data for a year, but then the data is archived and made available to everyone. So far, more than 8,700 scientific papers have been published using Hubble data. Every year more than 1,000 observation proposals are submitted by astronomers from all over the world, and around 200 are selected.

## BEGINNINGS

The development of the Space Shuttle in the 1970s revolutionized space telescope design, making it possible to use astronauts to aid the delivery of telescopes to orbit and allow mechanisms to be serviced, failed components replaced, and new and more up-to-date instruments mounted in the focal plane. Not only could a telescope start out as a state-of-the-art design, it could remain so throughout its fifteen-year planned lifetime.

The NASA design study for the Hubble Space Telescope commenced in 1973 and the European Space Agency joined the project in 1975. Construction and assembly took nearly a decade, and the whole spacecraft was finished in 1985. Launch was scheduled for 1986, but the Challenger accident on January 28th—where the seven crew members of the STS-51-L mission were killed when the shuttle disintegrated just 73 seconds after take off—and the subsequent Shuttle redesign put back the Kennedy Space Center launch until April 24th 1990. "First light," a momentous time for any telescope, occurred on May 20th 1990. Very quickly it was realized that there was a flaw in the main mirror that introduced spherical aberration and blurred the stellar images. Corrective optics,

PAGE 8–9
**Crowd of old stars**
The core of the star cluster Omega Centauri is crowded with stars. Most are yellow like our Sun. The red dots are larger, cooler, older stars. Some older stars appear blue. These have merged with other stars, boosting their energy production rate and turning them blue.

LEFT
**Atlantis stands ready**
On launch pad 39A at NASA's Kennedy Space Center in Florida, November 2009, Atlantis is ready to launch. The crew will service Hubble for the fifth and final time expanding its capabilities and extending its operational lifespan to at least 2014.

rather like a huge contact lens, was shipped up to the telescope on the first service mission in December 1993. A new wide field and planetary camera was also installed and sharp images started to be produced. In February 1997 seven astronauts flew to the Hubble Space Telescope and replaced two of the four focal plane instruments. The first replacement, the Near Infrared Camera and Multi-Object Spectrometer (NICMOS) had detectors cooled by evaporating solid nitrogen. Infrared radiation passes through interstellar dust more efficiently than visual radiation and so regions where new stars and planets were being formed could be probed in more detail. The second new instrument, the Space Telescope Imaging Spectrograph, could take detailed spectra of 500 places across a specific astronomical region. These spectra were then used to estimate the chemical composition, temperature, and relative velocity of the regions in question.

Six gyroscopes on board the HST help it maintain precision pointing. These were all replaced on the third servicing mission in mid-December 1999. A new central computer and new Fine Guidance Sensor replaced older, worn, and less sensitive versions. March 2002 saw astronauts back again. Another Wide Field and Planetary Camera was introduced.

By doing this the field of view (the area of sky being imaged) was doubled, and the speed at which data could be collected was increased by a factor of ten. As time passed, the telescope simply got better and better.

## THE FUTURE OF HUBBLE

The final trip to Hubble took place in May 2009. The visit ensured that HST works even better than before and at least to the year 2013. Astronauts installed two new instruments; the Cosmic Origins Spectrograph that works in ultra-violet light, and the Wide Field Camera 3 which has a resolution and field of view much greater than previous instruments. The ability to be continually upgraded and improved is one of the great advantages of the Hubble Space Telescope. Now rejuvenated, it continues to transfer about 120 gigabytes of data to Earth every week.

What is special about the year 2013? Well, space is an extremely harsh environment and space instrumentation does not last forever. By 2013 the Hubble Space Telescope will have been working for 23 years. It will be time for a replacement and this is known as the Next Generation Space

Telescope (NGST). 2016 is the year scheduled for the launch of the James Webb Space Telescope (the NGST being named after a former NASA administrator). The JWST is massive and is funded by the USA, Europe, and Canada. Instead of an HST main mirror 8 feet (2.5 m) across, the JWST will have a mirror that is 21 feet (6.5 m) in diameter, giving it seven times more radiation collecting area. The JWST mirror is made of eighteen segments that will be gently unfolded and adjusted to shape when in orbit. The JWST will not travel around Earth fifteen times a day as Hubble does; it will orbit around a position in space known as the second Lagrangian point, this being 930,000 miles from Earth, way beyond the Moon's orbit and diametrically opposite to the Sun in the sky. Needless to say, when in orbit, the JWST is completely beyond the reach of manned spacecraft. If anything goes wrong, no shuttle mission could be sent to fix it. The mirror and instrumentation will be cold and shielded from solar and terrestrial radiation by an enormous baffle about the size of a tennis court. The JWST is optimized for infrared observation and will have four science instruments on board that will concentrate on taking images and spectra. The plan is to probe the primordial early days of galactic evolution and to investigate the mechanisms responsible for star and planetary formation. After the hopefully successful launch and deployment of the JWST in 2016, the HST will most likely lose most of its funding and be "de-orbited"—its small station-keeping rocket systems will drive it into the Earth's upper atmosphere where it will burn up in a startling fireball.

The Hubble Space Telescope has been a huge success. There has been drama; corrective optics had to be supplied in the early days. There has been much bravery; traveling in the Space Shuttle, and working in low Earth orbit is not for the fainthearted. There have been breakdowns and instruments have needed to be replaced. But dedication and teamwork has triumphed, and the astronomical productivity has been enviable. When we look back on the history of the telescopic investigation of our Universe, the Hubble Space Telescope will always be remembered with pride. This book is a record of Hubble's momentous achievements, and will take you on a journey through time and space to view our wonderful, mysterious, and breathtakingly beautiful Universe through the eyes of this extraordinary telescope.

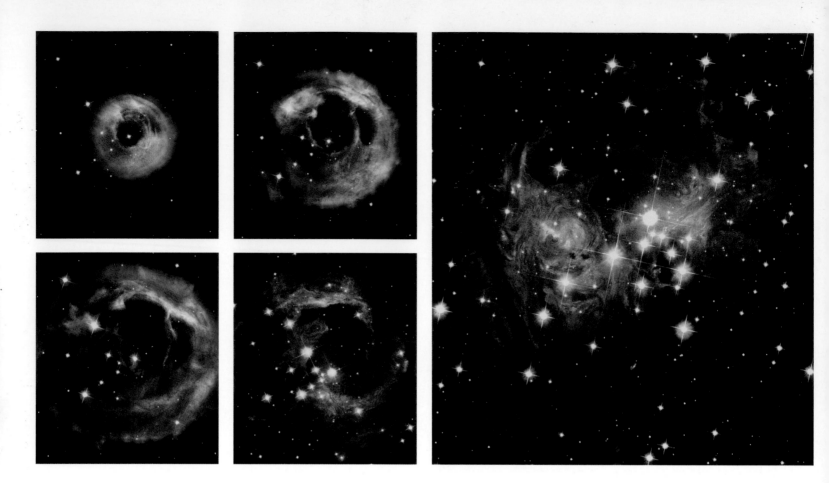

# Introduction

Before we look at the Universe through the HST's supersensitive eyes, let's set the scene. What, broadly speaking, is this Universe of ours like? One thing for sure is that it is vast—unimaginably vast. The Earth, Moon, planets, Sun, and stars are nothing but tiny specks of matter floating in an unfathomable immensity of space—minuscule, insignificant plankton floating in an infinitely deep cosmic ocean.

When we cast our eyes up to the night sky and see a glittering firmament of twinkling stars set in the velvety blackness that is space, we are looking out at one little corner of this ocean, of this Universe of ours.

Dominating the sky by night is the silvery Moon, Earth's closest companion in space—its only natural satellite. An airless, heavily cratered ball of rock, it is the only other world that human beings have set foot on—yet.

Next come the ultrabright stars that seem to wander through the heavens. But they are not stars at all: they are planets. What an extraordinary collection of bodies the planets are. They are all quite different from the planet we know best—planet Earth. Closer to the Sun, Mercury and Venus are oven-hot, while Mars farther out is cold, but might once have been warmer and supported some kind of life. Farther out still are gigantic Jupiter and other gas giants. And farthest out, are the dwarf planets, such as Pluto and Eris, two small ice worlds.

Dominating the sky by day is the golden orb of the Sun, which brings warmth, light, and life to Earth. It also dominates

near space with its powerful gravity, keeping the planets and a host of smaller bodies circling round it. All these bodies make up the Sun's family, or solar system.

The Sun is quite a different body from the planets: it is a huge globe of searing hot gas. It is our local star, just like the myriad other stars in the heavens but very much closer. The other stars, those pinpricks of light in the night sky, are so far away that their light takes years to reach us on Earth (light from the Sun takes just over 8 minutes). Astronomers use the distance light travels in a year (about 6 trillion miles/ 9.5 trillion km) as a measure for expressing distances in space. They call it the light-year. It is the unit we use throughout the book.

The Sun is a very ordinary star, of about average size (nearly 1 million miles/1.6 million km across) and average energy output. There are stars that are very much bigger and more energetic, and others that are smaller and emit much less energy.

Since the dawn of astronomy at least five millennia ago, stargazers have used patterns made by the bright stars to guide them across the night sky. These patterns are the constellations. Astronomers use Latin names for the constellations, which refer to figures ancient stargazers thought they could see in the patterns of stars. A few constellations live up to their names (Leo, the Lion; Scorpius, the Scorpion; Cygnus, the Swan), but most don't.

All stars are born in great billowing clouds of gas that occupy the space between the stars. After shining steadily for

ABOVE
**Echoes of light**
Astronomers are extremely lucky that the HST was in orbit at the time when the eruptive variable star V838 Monocerotis was seen to outburst in January 2002. This unusual star, 20,000 light-years away, suddenly became 600,000 times more luminous than our Sun. The red supergiant star in the center gave off a pulse of light which has been traveling through the dust surrounding the stars, revealing different regions as it moves outwards. The dust is thought to have been expelled in a previous explosion. Hubble has returned its gaze to the star several times since 2002. The four small images show the star's changing appearance from May 2002 to October 2004, and the larger image is from September 2006.

LEFT

**Ring of dark matter**
Regions containing dark matter cannot be seen, but their gravitational fields influence the path that light takes as it passes through. In November 2004 the HST imaged the cluster of galaxies known as Cl 0024+17, some 5 billion light years away. The image shows how the gravity of the dark matter in the cluster distorts the light coming from more distant background galaxies. The ghostly ring, some 2.6 million light-years across, was probably produced by a collision between the cluster of galaxies and another cluster some 1 to 2 billion years ago. The ring shape is produced because the colliding clusters were moving along the line of sight.

millions or billions of years, stars begin to die. They may exit the Universe relatively quietly, as the Sun will eventually, or blow themselves apart in a fantastic supernova explosion. The end products of their death throes will be superdense bodies like white dwarfs and neutron stars, or the most awesome objects we know in the Universe—black holes.

The stars we see in the night sky may lie many thousands of light-years away, but they are still close neighbors in the Universe. They all belong to a great star island in space—a galaxy. The Universe is made up of at least 125 billion galaxies, separated by virtually empty space.

Our galaxy, called the Milky Way or just the Galaxy, probably contains at least 500 billion stars. It measures some 100,000 light-years across and has a disk shape with spiral arms. Many galaxies are like it, but others are elliptical in shape or have no regular shape at all. We can see just three galaxies at night with the naked eye. They are the two Magellanic Clouds in the far southern skies and the Andromeda Galaxy in northern skies.

Some galaxies are extraordinary, pumping out much more energy into the Universe than usual, particularly at radio wavelengths. Called active galaxies, they include enigmatic bodies such as quasars and blazars. Black holes seem to be the engines that generate their exceptional power.

Just as stars group together to form great stellar island galaxies, so galaxies themselves group together to form clusters. Our own Galaxy is part of a relatively small group of about 40 galaxies. But we know of clusters containing thousands of galaxies. In their turn, even clusters group together to form superclusters. And on the largest scale, it is strings of superclusters, interspersed with empty voids, that make up the Universe.

How can we put such an enormous Universe in perspective? With great difficulty—but we can try. Let's suppose we have been able to build an interstellar and intergalactic starship, capable of traveling at the speed of light, and take an incredible journey into space. Setting off from Earth, we would reach the Moon in 1½ seconds and flash pass Venus in 2½ minutes. Soon we would be leaving the Sun behind, heading for the distant dwarf planet Pluto. We would reach this tiny world in 5½ hours. But it would take a year or so before we escaped completely from the gravitational influence of the Sun and left the solar system. Now traveling in interstellar space, we wouldn't reach even the nearest star (Proxima Centauri) for more than 4 years.

To explore our Galaxy would require flight times measured in tens of thousands of years—27,000 years to reach the Galaxy's center and almost twice as long again to reach its farthest edge. To make a visit to our galactic neighbor, the Andromeda Galaxy, we would have a journey time of 2.9 million years. And to reach the farthest objects we can see in the Universe, we would have to journey for at least 13 billion years. This is nearly as long as the Universe has existed.

# 1 | Stars in the Firmament

**Dark, dusty clouds are the birthplace of stars**

**ABOVE: Brilliant, young stars**
Newly born stars embedded within an outer region of the huge Eagle Nebula shine brightly. Ultra-violet radiation emitted by the stars causes the surrounding star-forming nebula to glow. These brilliant stars are part of the open star cluster NGC 6611 which formed from the nebula material.

**INSET LEFT: Seeding interstellar space**
Jets of gas stream from an enigmatic object known as He2-90 in Centaurus. It seems to be a close pair of dying stars masquerading as a single youngster—most stars emit jets in their youth. Dying stars add substance to the interstellar medium by shrugging off large amounts of gas and dust.

**INSET RIGHT: Clustering together**
A dazzling cluster of stars in one of our galactic neighbors, the Small Magellanic Cloud. These relatively young stars in the cluster (NGC 265) were born from the same cloud of interstellar gas and dust but will drift apart. The image was taken with the HST's Advanced Camera for Surveys.

# THE UNIVERSE OF GAS AND DUST

In every direction we look in the night sky, we see stars. If we were very patient and very meticulous, we could count around 5,000 stars visible to the naked eye. Looking through binoculars, we would see thousands more stars, and through a telescope, stars in their billions.

Even a casual glance reveals that stars are not all the same. Some are bright, others dim; most are white, others colored. Brightness and color are two factors that make stars different. But when we investigate stars closely with telescopes and analyze their faint light, we find that they also differ in many other ways—in temperature, mass, speed, magnetism, composition, age, and so on. In particular, other stars can differ remarkably from the star we know best—our local neighborhood star, the Sun.

So many kinds of stars. Tiny stellar dwarfs that glimmer feebly, like celestial glowworms; supergiant stars thousands of times bigger and millions of times more brilliant that stand out like cosmic beacons; stars seemingly with a dimmer switch that makes them vary in brightness; newborn stars with the exuberance of youth; ancient stars racing toward celestial Armageddon.

There are so many different species in the stellar zoo. But in studying these disparate bodies, astronomers discover that each kind represents a different stage of stellar evolution. From this, they can piece together the life cycle of a typical star, from its birth, through its youth, into middle age, and finally to its death. Astronomers have to do things this way for they can't follow a single star through all its evolutionary stages, which, of course, would take billions of years.

The story of the stars begins not with the stars themselves, but in the vast space that exists among the stars. We tend to think of this space as being empty, but it isn't—quite. Scattered about interstellar space are atoms of hydrogen gas and tiny grains of dust. On average, there are a few dozen hydrogen atoms and the odd speck of dust in a volume of space the size of a soda can. This means that interstellar space is millions of times more empty than the most perfect vacuum that scientists can achieve on Earth.

The tenuous mixture of gas and dust among the stars is known as the interstellar medium, and its typical composition is called the cosmic abundance of the elements. Hydrogen makes up about 75 percent of the interstellar medium, and helium makes up about 23 percent. The remaining 2 percent is made up of traces of heavier elements, which astronomers rather confusingly call "metals," though this term also encompasses elements such as carbon and oxygen, which scientists consider non-metals.

Carbon and oxygen are among the most abundant of these "metals," carbon being the primary constituent in interstellar dust grains. Nitrogen and iron are relatively common as well. There are also sprinklings of all the other heavier chemical elements—from arsenic to tin, lead to gold.

Where exactly do all of these elements come from? The hydrogen and helium are primordial—they were formed at the very beginning of the Universe. All the other elements, however, have been forged in the interior of stars. Ordinary stars such as the Sun produce carbon and oxygen as they enter old age. Supergiant stars produce iron. But all heavier elements are produced in the supernova explosions that mark the death of supergiants.

All these elements are puffed or blasted into the interstellar medium when a star dies and subtly changes its composition.

ABOVE
**Starbirth pillar**
This huge pillar of gas and dust is in the Carina Nebula, a vast star-forming region. Within the pillar but hidden from view are newly-forming stars. The pillar is 3 light-years long and lit by hot nearby stars, just off the top of the image. Winds from these stars are sculpting the pillar.

RIGHT (MAIN IMAGE)
**Beyond the dust**
Hubble's Wide Field Camera 3 took these two images of R136, a cluster of young stars, in October 2009. Only a few million years old, the cluster nestles in the 30 Doradus Nebula within the nearby galaxy, the Large Magellanic Cloud. The view inset below is in ultra-violet and visible light; the stars are blue and the hydrogen red. The main image is in the infrared and sees through the dusty nebula to reveal stars not normally visible.

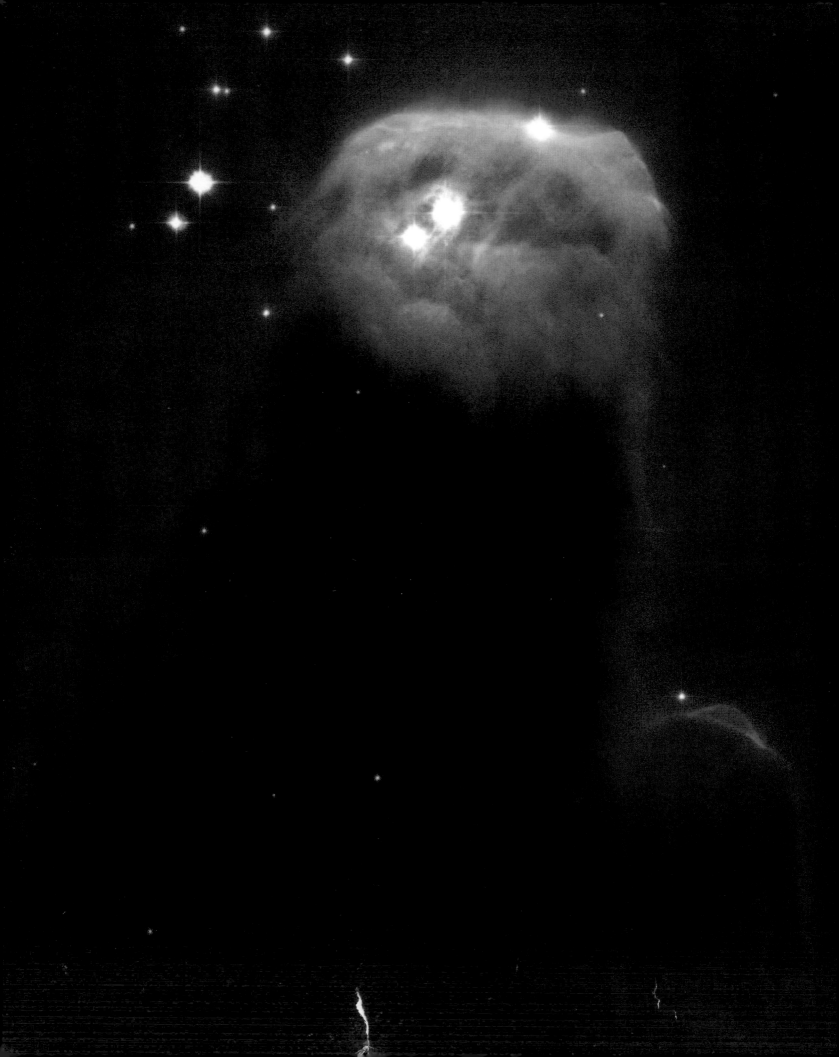

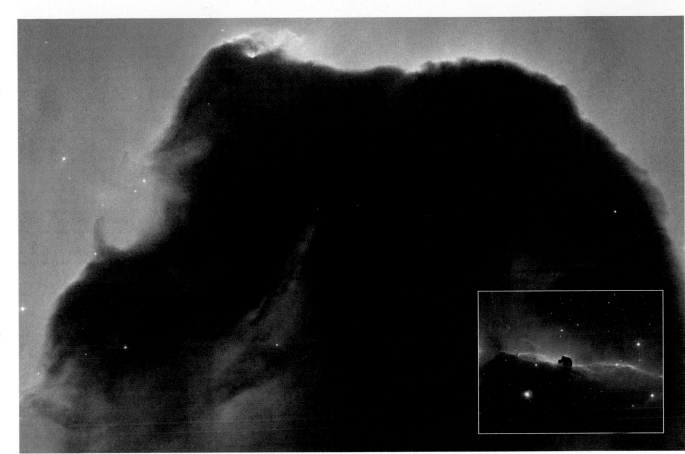

## MOLECULAR CLOUDS

In many regions of interstellar space, the gas and dust becomes relatively dense, though still nowhere near as dense as air. It then forms into vast cloud-like masses. Typically, these clouds are very still and cold, around minus 430 degrees Fahrenheit (–260°C). Under these conditions, the atoms of the elements present join to form molecules, which is why these cool, dense regions are called molecular clouds.

The most common molecule in the clouds is, of course, hydrogen. Scientists have identified many other interstellar molecules, usually by investigations done at infrared, microwave and radio wavelengths. Every molecule has a characteristic wavelength "signature" that identifies it.

Among the other molecules found in molecular clouds are water, ammonia, hydrogen sulfide, formic acid, methanol, and glycine. The presence of glycine is particularly interesting because this compound is an organic substance called an amino acid. And amino acids are the building blocks of proteins, which are essential to life as we know it. So the presence of glycine and the other organic chemicals in interstellar space suggests that there may be forms of life elsewhere in the Universe.

## THE DARK SIDE

Dark molecular clouds lurk throughout the Galaxy. Most merge into the inky blackness of space. But some become visible when they blot out the light of stars or glowing gas behind them. We call these molecular clouds dark nebulae—from the Latin word for clouds.

We can see two huge dark nebulae with the naked eye. One, called the Cygnus Rift, lies in the constellation Cygnus. The other, appropriately named the Coal Sack, lies in the far southern constellation Crux, the famous Southern Cross. Best-known among dark nebulae visible with telescopes is the Horsehead in the constellation Orion. It really does look just like a horse's head and flowing mane.

## ALL LIT UP

Sometimes the clouds of interstellar gas are lit up by nearby stars and become visible as bright nebulae. They are lit in one of two ways: from the outside or the inside.

When there are stars near the Earth side of a nebula, it reflects their light toward us and is called a reflection nebula. When the stars are embedded in the nebula, radiation from them excites the gas molecules, making them emit their own radiation. We then call it an emission nebula. An emission nebula is characteristically red, as this is the wavelength of light emitted by excited hydrogen atoms.

There are many bright nebulae visible in the heavens. These are predominantly emission nebulae, but they also usually feature smaller regions that glow by reflection. Brightest by far and visible to the naked eye is the Orion Nebula (profiled on page 24).

When you look through a telescope, hundreds of bright nebulae swim into view. Many have distinctive shapes that have earned them popular names, such as the Eagle Nebula in Serpens, the Lagoon Nebula in Sagittarius, and the North America Nebula in Cygnus.

# PROFILING:
# THE ORION NEBULA

RIGHT
**The Trapezium**
In this infrared view of the Orion Nebula, the stars of the multiple system we know as Theta Orionis, or the Trapezium, shine brilliantly. The picture also reveals dozens of feebly glowing brown dwarfs—objects too small and too cool to achieve true stardom.

RIGHT
**The Orion Nebula**
The bright patch we see in the sky as the Sword of Orion is seen as a turbulent star-formation region when observed through high-power telescopes. In one of the most detailed astronomical images ever produced, the HST has revealed a tapestry of star formation, from dense pillars of gas and dust to the hot, massive new-born stars.

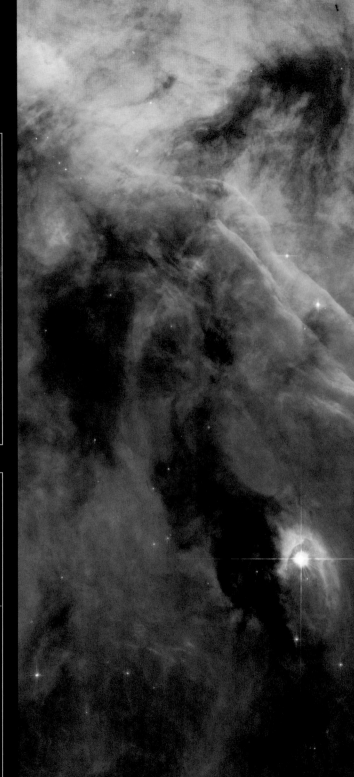

Orion is perhaps the finest constellation in all of the heavens. Spanning the celestial equator, it is visible to stargazers anywhere in the world. Orion is one of the few constellations that bears more than a passing resemblance to the figure it is named for. It depicts a mighty hunter from Greek mythology, in his left hand a shield and in his right a club, raised and ready to strike.

The individual stars that make up Orion are spectacular. The bright red supergiant star Betelgeuse marks his right shoulder, while brilliant white Rigel marks his left knee.

Defining Orion's Belt are three slightly less bright stars, and hanging from the belt is his sword.

The sword is marked by a bright, misty patch known as the Orion Nebula, or M42. Binoculars reveal greater detail, and with a small telescope we can make out individual stars. But long-exposure photographs in larger instruments are needed to bring out fully the nebula's incredible beauty.

Some 1,600 light-years away, M42 is the nearest bright nebula to Earth. It is what astronomers often call a blister

ABOVE
**Zooming into the Orion Nebula**
This panorama is a part of the Orion Nebula. It is one in a series of detailed views made from the HST's 2006 portrait (left), showing the region just below the center. The Trapezium stars, off the image to the upper left, illuminate the thick swirls and streams of gas.

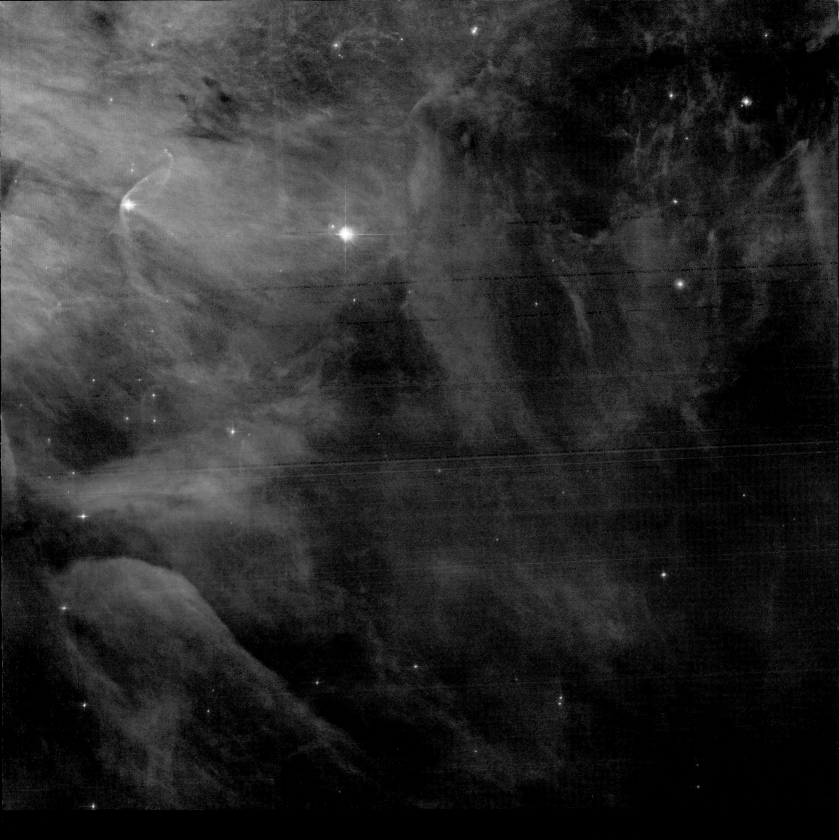

nebula—a region at the edge of a dark molecular cloud that has been lit up by the powerful ultraviolet light of nearby embedded stars.

In M42, the star in question is Theta Orionis. Actually, it is a multiple star system, known as the Trapezium because of the arrangement of its four stars. Only about 300,000 years old, these stars are in their infancy, and they are pumping out vast amounts of energy in the ultraviolet.

But this is not the only region in Orion where star formation is taking place. Much of the constellation is embedded in vast, billowing clouds of gas and dust. There are two particularly dense areas, known as the southern and northern molecular clouds.

M42 is part of the southern cloud, which nearly merges into the northern cloud around Zeta Orionis, the most southerly star in Orion's Belt. The outstanding feature of the northern cloud is the Horsehead Nebula, which is thrown into dramatic silhouette by the bright nebula beyond (see page 21).

# A STAR IS BORN

The cold, dark molecular clouds found in Orion and elsewhere in our Galaxy can stretch for hundreds of light-years. It is in such giant molecular clouds that stars are born. These clouds can remain relatively still and quiescent for hundreds of millions of years. Then something disturbs them, triggering a chain of events that will ultimately lead them to spawn new generations of stars.

Exactly what triggers a disturbance still puzzles astronomers. It could be the shock wave from a nearby supernova explosion, a collision with another giant molecular cloud, or a galactic pressure surge. But the end result is the same: parts of the cloud become so dense that gravity becomes the most dominant force, and they start to collapse.

## BARNARD AND BOK

The collapsing regions in a molecular cloud may contain enough gas and dust to create hundreds of stars. Such regions are known as Barnard Objects, after the U.S. astronomer Edward Barnard. Typically tens of light-years across, they are the kind of dark nebulae that we often see obscuring the light of background stars. We see other, smaller collapsing regions as little black bubbles against a background of stars or a bright nebula. They are called Bok Globules, after the Dutch astronomer Bart Bok. Only one or two light-years across, these contain enough mass to make about 20 to 200 sunlike stars.

As both Barnard Objects and Bok Globules collapse, clumps of matter within them collapse as well. So the original cloud fragments successively into smaller and smaller clumps. And it is from the smallest clumps that individual stars form.

## GRAVITY RULES

In the center (core) of a collapsing clump, concentrations of matter build up. The core is a star in embryo—a protostar. The denser the protostar becomes, the stronger its gravity will be, and it attracts more surrounding matter. As gravity tugs at it, matter speeds up; and gravitational energy converts into kinetic energy—the energy of motion.

When this moving matter suddenly reaches the core, it collides with the matter already there and stops. Its energy immediately converts into another form—heat. So, as the protostar accumulates more and more material, its temperature rises.

## IN A SPIN

The protostar is not stationary. The original giant molecular cloud rotates, and it imparts this rotation to the collapsing clump and the protostar within. The spinning mass then flattens out. Over time a thick disk, or ring, of gas and dust builds up around the protostar, while its temperature rises rapidly. It is already giving off radiation, mainly at infrared wavelengths.

As more and more matter rains down, temperatures inside the protostar soar to millions of degrees, and pressures reach millions of atmospheres. Under these exotic conditions, the nuclei (centers) of hydrogen atoms are forced together, and they fuse (join) to produce helium. This process of nuclear fusion produces huge amounts of energy (see page 132). The protostar springs to life and begins pouring out light, heat and other radiation into the Universe. A new star is born.

ABOVE
**Colorful jets**
At the center of this HST image of Herbig-Haro object 32 (HH 32) is a young star blasting jets of matter into space in polar jets. One jet (at the top) can be seen plowing into interstellar gas and making it glow in the light of hydrogen atoms (green) and sulfur ions (blue). The jet streaming in the opposite direction is mostly obscured by dust.

LEFT
**Bok globule**
A dense dark knot of gas and dust called a Bok globule is seen silhouetted against the background of the Carina Nebula. It is nicknamed "the caterpillar" because of its resemblance to the insect. Once a Bok globule has accumulated enough gas and dust it can create stars, but some dissipate before they reach that stage.

BELOW
**Enigmatic "waterfall"**
Both polar jets from a protostar can be seen slamming into interstellar gas in this VLT (Very Large Telescope) image of HH 34. The nature of the prominent waterfall-like stream of light is a mystery.

RIGHT
**Double bubble**
A cocoon of gas and dust surrounds a small cluster of young, hot stars in the Large Megallanic Cloud. This "double bubble" lies inside the larger nebula, DEM L106, one of many star-forming regions in the galaxy.

## REACHING EQUILIBRIUM

Once the nuclear furnace of a star fires up, radiation floods out of it. Until this point, the star has been contracting under gravity. But now radiation pressure starts pushing in the other direction, causing the star to expand. It continues to expand until it reaches an equilibrium point, at which the outward push of the radiation balances the inward pull of gravity. At that point, the newborn star reaches the steady state it will assume for many millions—or even billions—of years. It becomes what astronomers call a main sequence star (see page 35).

In reality, however, things are not that simple. It takes a star quite a while to reach the equilibrium point, typically about 10 million years. When it first starts to expand, battling against gravity, it expands too far and blows material off into space. Then gravity takes charge for a while, forcing the star to contract again. Soon the cycle starts all over again—expansion, mass loss, contraction; expansion, mass loss, contraction. Slowly, the amplitude of these oscillations subsides, until the star finally achieves a state of equilibrium.

The cycle of expansion and contraction causes the star's output of light to fluctuate erratically. This period of a star's life is called the T Tauri phase, after the first star of its kind that was discovered.

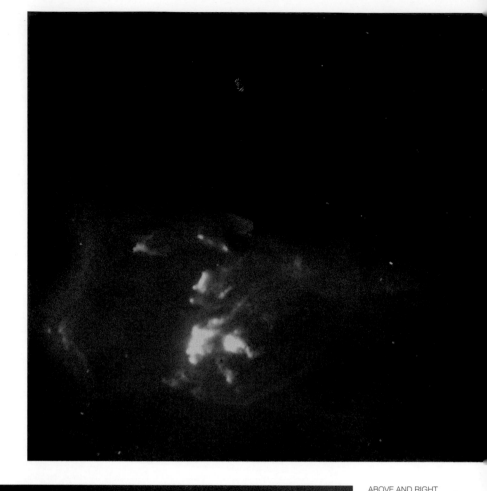

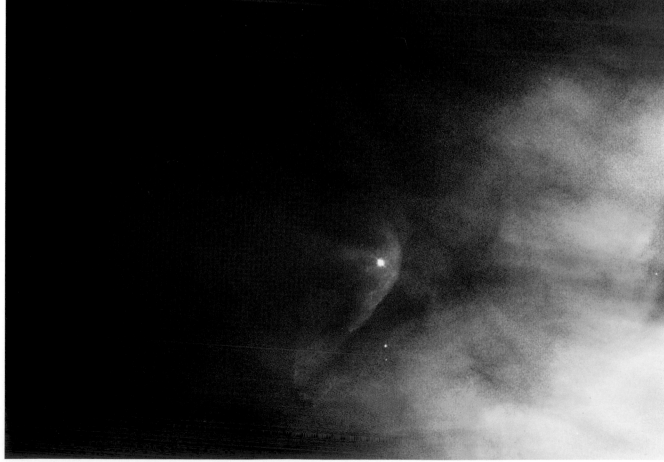

ABOVE AND RIGHT
**Jet exhausts**
Jets are common "exhaust" products of star formation, revealed when they ram into surrounding gas and dust. In these images, the central stars are hidden within masses of infalling material. Typically, the twin jets from a fledgling star span a region of space about two light-years across.

LEFT
**Making waves**
Gas streaming from the very young star LL Ori collides violently with the tenuous interstellar medium, creating a "bow shock" effect around it.

## JET-PROPELLED

In these early days of a star's life, when its nuclear furnace is stoking up, it is still surrounded by a veil of gas and dust and a thick disk of material. It also gives off streams of particles as a kind of stellar "wind" that starts to blow away the surrounding material. This can't happen around the equatorial regions because of the surrounding disk. But it can and does happen at the poles, where matter emerges as powerful jets. The jets stream out into space and collide with interstellar gases, creating "knots" of glowing gas known as Herbig-Haro objects. The jet phase of a star's life is fleeting, lasting only a few thousand years.

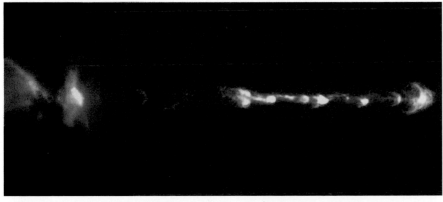

CoKu Tau1

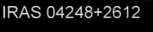

IRAS 04248+2612

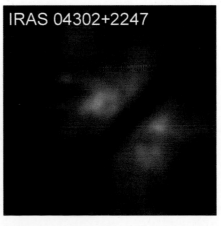

IRAS 04302+2247

LEFT
**Planets in the making**
Dark, dusty disks around embryonic stars show up in these infrared images. When the stars begin shining steadily, the disks will probably remain, providing raw materials for future planetary systems.

Serpens (the Serpent) is an oddity among the constellations because it is split in two: Serpens Caput—the serpent's head—and Serpens Cauda—the serpent's tail. In between is the constellation Ophiuchus, the Serpent-Bearer.

Ophiuchus, like Serpens itself, is not an easy constellation to identify because it has no particularly bright stars. One of its claims to fame is what it's not—it's not considered by astrologers to be a constellation of the zodiac. And it should be, because the Sun spends longer passing through Ophiuchus than it does passing through Scorpius, which is a zodiac constellation. The omission of Ophiuchus as a star sign (or zodiac constellation) is just one serious weakness in the case for astrology.

But, back to Serpens. Serpens Caput has a fine globular cluster, M15. Serpens Cauda has M16, which is a bright nebula vaguely shaped like a bird with outspread wings. Known as the Eagle Nebula, M16 was the subject of one of the most dramatic images the Hubble Space Telescope (HST) has ever taken.

The HST science team called the image "The pillars of creation." It shows dark columns of gas in which stars are being born. The columns, or pillars, are etched and silhouetted by the light of young, hot, massive stars beyond. The pillar on the left is about seven light-years long.

The finger-like protrusions at the top of the pillars are dense regions that probably contain newborn stars or protostars. Termed EGGs (evaporating gaseous globules), they have been revealed because intense ultraviolet radiation from hidden massive stars has blown less dense gas away. Eventually the radiation will blow away the gas in the EGGs as well, revealing the star inside for the first time.

ABOVE
**Sculpted pillar**
**This detail of an Eagle Nebula pillar was released by the HST team in April 2005. Light from nearby bright, hot, young stars is sculpting the cloud of gas and dust into intricate forms and causing the gas to glow.**

**Stellar Stalagmites**
The eerie pillars of dark, dense gas in the heart of the Eagle Nebula are incubators for new stars. They protrude from the wall of a dark molecular cloud rather like stalagmites from a cave floor.

**Stellar spire**
A billowing tower of cold gas and dust rises from the Eagle Nebula. At 9.5 light-years or 57 trillion miles high, it is about twice the distance from our Sun to the next nearest star. The bumps and fingers of material appear to be regions where new stars would form. The blue coloring is glowing oxygen, and the red is glowing hydrogen.

# WHAT STARS ARE LIKE

When a molecular cloud collapses, the stars that form vary widely in mass. A star's mass determines how long it will shine. The Sun's mass is described as one solar mass and the mass of other stars is described in relation to this; either a multiple, or a fraction of it. The average star is about one seventh the Sun's mass. Stars with very little mass, less than a fifteenth of the Sun's don't exist because they don't accumulate enough mass to generate the heat and pressure necessary to trigger nuclear fusion. These "failed stars" glow a dull brown and we call them brown dwarfs.

A collapsing cloud tends to produce relatively few really massive stars. But they can be huge, with up to 50 times the mass of the Sun and 20 times its diameter. Really massive stars don't exist because they generate so much energy at the protostar stage that they blow themselves apart.

## MASS AND TEMPERATURE

The mass of a main sequence star determines its temperature. The low mass stars convert their hydrogen to helium relatively slowly and have a low surface temperature, maybe below 5,500 degrees Fahrenheit (3,000°C). Stars with the Sun's mass are about twice as hot, while the most massive giant stars have temperatures beyond 55,000 degrees Fahrenheit (30,000°C).

Allied to a star's temperature is its brightness. Here we are talking about a star's luminosity and not its apparent brightness as we perceive it from Earth. As you might expect, the least massive are the dimmest—they may be only 1/100,000,000th as bright as the Sun. And the most massive stars are the brightest—up to 100,000 times brighter than the Sun. This brightness is the star's luminosity, which is compared to that of the Sun.

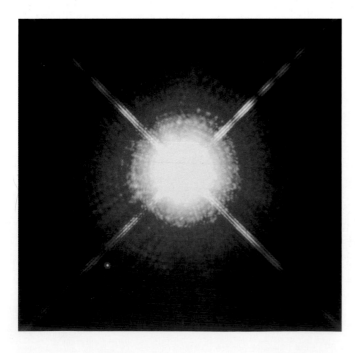

## TEMPERATURE AND COLOR

The temperature of a star also determines its color. Think of an iron plunged into a fire: As its temperature rises, it gradually changes color from dull to bright red, then orange, yellow and finally brilliant white. And it is a similar story with stars.

Low-mass, small, cool stars that glow red are called red dwarfs. More massive, larger, and hotter stars like the Sun glow yellow; these are the yellow dwarfs. The most massive and largest, hottest stars glow blue–white; we call them blue giants.

## THE INFORMATIVE SPECTRUM

When we see a rainbow in the sky, we are looking at a range of different colors, from violet to red. Mixed together, these colors make up white sunlight. The spread of color, or spectrum, is a kind of fanning out of the different wavelengths in light because we perceive different wavelengths as different colors. Violet light has the shortest wavelength, red the longest. Scientists create a spectrum artificially by passing sunlight through an instrument called a spectroscope or spectrograph. The technique is called spectroscopy. The light from all stars can be split into a spectrum in a similar way.

Spectroscopy is one of the most powerful tools in astronomy, since a star's spectrum holds the key to its identity. When you closely examine the rainbow-like spectrum of any star (including the Sun), you notice a series of dark lines. These are called absorption lines, because they are produced when certain wavelengths of starlight are absorbed as they pass through the star's atmosphere. Where these lines appear in the spectrum depends on what elements are present in the atmosphere.

Every element absorbs different wavelengths and gives rise to a characteristic set of spectral lines. So by studying the spectrum, astronomers can work out the composition of a star and much more besides. They can also estimate the star's temperature, density and magnetism; tell how fast it is spinning; and determine whether it is traveling toward or away from us.

## BE A FINE GUY

All stars produce a spectrum with dark absorption lines, but each kind of star displays a slightly different and characteristic spectrum, depending on its color (and thus its temperature).

By this means, astronomers divide stars into 10 main groups, or spectral classes, designated O, B, A, F, G, K, M, R, N, and S. (Difficult to remember? Try the once-heard, never-forgotten mnemonic "Oh, Be a Fine Guy, Kiss Me Right Now, Sweetie!")

O and B stars are the hottest; S stars are the coolest. The cooler the star, the more lines appear in its spectrum.

# THE H-R DIAGRAM

With such a variety of different stars—bright, dim; hot, cool; blue, red; large, small—how is it possible to classify them? To answer this question, we must go back to two of the leading astronomers from the early part of last century: Ejner Hertzsprung of Denmark and Henry Norris Russell of the United States.

Working independently, they discovered that there is a significant relationship between the luminosity of a star and its spectral class/temperature. When they plotted luminosity against spectral class for a variety of stars, they produced a very interesting graph, which is now called the Hertzsprung-Russell (H-R) diagram.

## ON THE MAIN SEQUENCE

On the H-R diagram, stars are not scattered about randomly as you might expect. Instead, they fall into a number of fairly distinct groups. The majority lie within an elongated, S-shaped, diagonal band called the main sequence.

Dim, cool red dwarfs, such as Barnard's Star and Proxima Centauri, are found on the lower right of the main sequence. Bright, hot blue giants, such as Spica and Regulus, are found on the upper left. The Sun, a yellow dwarf, is nearly halfway up the main sequence. In effect, the position of a star on the main sequence depends on its mass. The more massive a star, the hotter it is, because it burns more fuel (hydrogen), faster.

## COOL GIANTS AND HOT DWARFS

Stars are also found in three other main regions on the H-R diagram. Giant stars like Aldebaran and Arcturus are found above the main sequence; they are bright but cool. Supergiants such as Canopus and Antares lie near the top of the diagram; they may be hot and blue or cool and red. The very hot but very tiny bodies we call white dwarfs are at the foot of the diagram.

**RIGHT**
**Sparkling jewels**
Fourteen massive stars are on the verge of exploding as supernovae in this jewellery box full of sparkling stars. Each is a red supergiant about 20 times as massive as the Sun. They are inside the bluish cluster in the center of the image.

## LIFE CYCLE

The H-R diagram reflects a frozen moment in the lives of stars we can see in the heavens today. If in a few billion years time we were to plot an H-R diagram again, it would look completely different.

Stars remain on the main sequence for most of their productive lives—that is, when they are burning hydrogen in their cores. But not all stars are the same. Red and yellow dwarf stars remain on the main sequence for billions of years, but hot blue giants pay only a fleeting visit, measured in a few million years.

When stars have used up all their hydrogen, they begin to die and leave the main sequence. The biggest ones become supergiants, the smallest, red giants. Eventually, supergiants blow themselves to bits and disappear from the diagram; some even disappear from the visible Universe (as black holes). Red giants shrink and become very hot, turning into white dwarfs. We follow the often spectacular death throes of stars in the next chapter.

ABOVE
**Dim dwarfs**
Within the ancient globular star cluster NGC 6397 (main image), the HST's Advanced Camera for Surveys has identified the faintest red dwarfs (circled in the lower right image) and the dimmest white dwarfs (circled in the upper right image).

BELOW
**The H-R Diagram**
A selection of familiar stars plotted on an H-R diagram. They tend to fall into four main groups. Stars in the prime of life ride the main sequence. Giant and supergiant stars lie above this diagonal band, white dwarfs below.

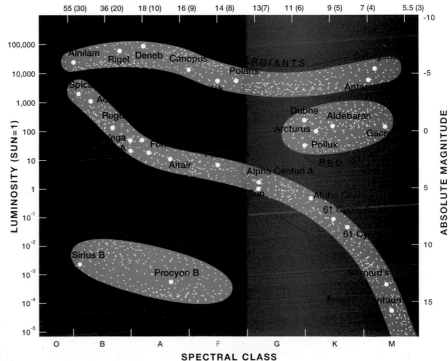

# CLUSTERING TOGETHER

Stars are born in the giant molecular clouds that exist between the stars, but they are rarely born one at a time. Usually, many clumps of matter condense within a cloud simultaneously, and each one may develop into a star. And if two or more stars are born close together, they may eventually travel as companions through space.

The Sun, of course, has no stellar companion and moves alone in space—except for its family of planets and other bodies. Roughly a third of stars lead solitary lives, but the rest travel with one or more "sibling" stars.

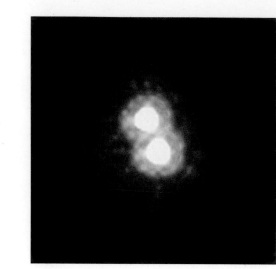

## BINARIES

Nearly half of all stars are double stars called binaries. In these systems, the two component stars orbit one another in a perpetual celestial waltz. Sometimes we can separate the components—see them separately—in a telescope. But often the components are so close together that we can identify them only by the shifts in their spectral lines as they move to and fro. In this case they are called spectroscopic binaries.

Binaries are not to be confused with optical doubles. These are pairs of stars that are actually far apart, but appear close together only because they happen to lie in the same direction in space as viewed from Earth.

## THE WINKING DEMON

A particularly interesting kind of binary is called an eclipsing binary. In this system, the two component stars circle round each other in our line of sight, and each star periodically passes in front of (eclipses) the other. The overall brightness of the system dips briefly with each eclipse, before regaining its former brilliance when the eclipse is over. Eclipsing binaries are a common kind of variable star—one that varies in brightness. The star Algol ("the winking demon") in Perseus is a classic example. Other kinds of variable stars, such as Cepheids, vary in brightness because of processes going on inside them (see page 105).

## OPEN CLUSTERS

Sometimes hundreds of stars are born close together and are companions for a while. There are many groups of stars like this in the heavens, called open clusters. The stars are three times closer together than the stars in the vicinity of the Sun.

We can see several open clusters with the naked eye. The easiest to spot is the Pleiades in the constellation Taurus. It is also called the Seven Sisters, because keen-sighted people may be able to see its seven brightest stars. There is also another open cluster in Taurus, called the Hyades. It surrounds (but is not part of) the noticeably orange star Aldebaran, which marks the eye of the bull.

## YOUNG UPSTARTS

Most open clusters are made up of young, hot stars. The Pleiades stars were born about 75 million years ago; those in the Jewel Box in Crux, only about seven million years ago. These clusters are the new kids on the block as far as the Universe is concerned, but they will not always stay together. Their gravitational ties are tenuous, and over time they will disperse and gain their independence.

LEFT
**Family portrait**
This infrared image shows a stellar family of a bright, massive mother star that is surrounded by fainter infants. Violent stellar winds released during the formative years of the mother star would have triggered star formation in some of the surrounding gas clouds.

ABOVE
**Heavyweight stars**
Two massive stars, known as Pismis 24-1, orbit one another. They and a third companion, here hidden in the glare of the other two, are estimated to have a combined mass of 200 times that of the Sun. They are amongst the heaviest stars known.

RIGHT
**Ripples in Tarantula**
The star cluster (lower right) in the image, Hodge 301, is located at the edge of the Tarantula Nebula in the Large Magellanic Cloud. It contains a mix of brilliant, massive stars, including several red supergiants that will soon explode as supernovae. Other stars in the cluster have already exploded, sending shock waves rippling through the nebula.

## GLOBULARS

In the constellation Hercules is an object designated M13. It is just visible to the naked eye, and can be clearly seen in binoculars. But a powerful telescope is needed to show it in detail. It proves to be a huge mass of stars packed tightly together into a globe shape and numbered in their hundreds of thousands.

M13 is a concentrated group of stars called a globular cluster. It is not unique. About 150 have been found in our Galaxy, and many more exist in others. The two most spectacular globular clusters in our Galaxy are found in far southern skies, and both are easily seen with the naked eye. Looking rather like hazy stars, they are so identified, as Omega Centauri and 47 Tucanae.

Omega Centauri, in the constellation Centaurus, is thought to contain at least a million stars. It is the biggest globular we know, measuring 180 light-years across. 47 Tucanae, in Tucana, may well have as many stars; it is slightly smaller but has a more concentrated central region.

## IN THE CENTER

Open clusters like the Pleiades are made up of young, hot stars and are found in the spiral arms of our Galaxy (see page 36). Some of these clusters lie quite close to us.

Globular clusters, however, are made up of old stars and are found in a spherical halo around the center of the galaxy. They do not take part in the Galaxy's general rotation, but pursue independent orbits around the galactic center. This takes them well above and below the plane of our Galaxy into the spherical region known as the halo.

Whereas the young and hot stars in an open cluster usually appear blue, globular cluster stars appear yellow because they are older and cooler. Astronomers reckon that globulars are typically about 10 billion years old, so they must have formed during or shortly after the formation of the Galaxy itself.

Studying these ancient stellar bundles helps astronomers to chart the early history of our Galaxy. Data collected by the HST plays a particularly crucial role in this study. Individual stars in the very heart of the globulars are resolved (viewed separately). Additionally, the HST has spotted globulars in other galaxies, some of which have proved to be much younger than those in our own.

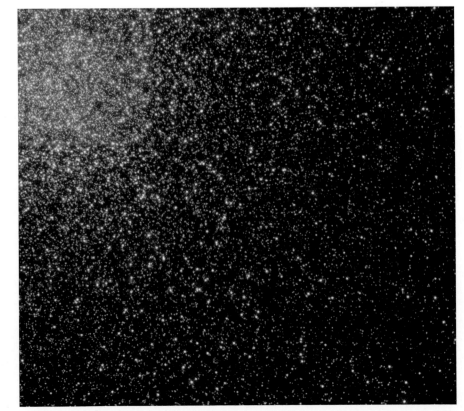

ABOVE
**Northern cluster**
Hundreds of thousands of individual stars are visible in this Hubble image of the globular cluster M13. It is one of the brightest globular clusters in the northern sky. It lies about 25,000 light-years away but is easily seen through a small telescope in the constellation of Hercules.

LEFT
**In the center**
This HST view shows about 35,000 stars in the central (upper left) region of 47 Tucanae. Most of the stars in the cluster formed about 10 billion years ago.

LEFT
**Ancient stars**
The stars in the globular cluster M15 look fresh and bright but they are some of the oldest in the Universe; they are roughly 13 billion years old. The densely-packed spherical cluster is in the outer reaches of the Milky Way and orbits the center of our Galaxy.

# 2 | Stellar Death and Destruction
Dying stars exit the Universe spectacularly

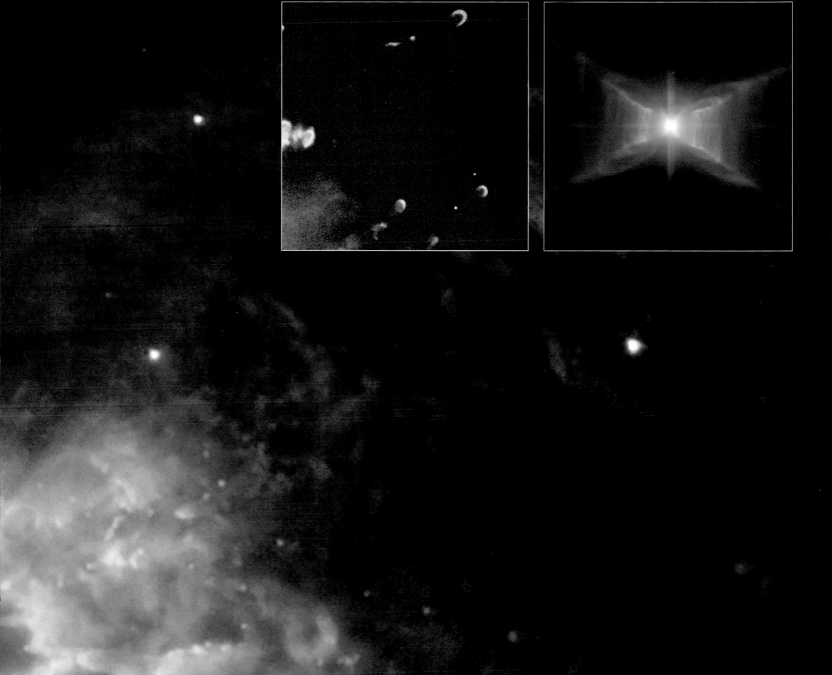

ABOVE
**Demise of a star**
The HST's Wide Field Planetary Camera 2 took this image of a dying star on February 6th 2007. The star is ending its life by casting off outer layers of gas. The remains of the star, now a white dwarf, is the dot in the blue central region of the colorful gas clouds known as NGC 2440.

INSET LEFT
**Cometary knots**
Tadpole-shaped blobs stream from the inner edge of the Helix Nebula in Aquarius. They are called cometary knots, though they have nothing to do with comets. The Helix Nebula is a close planetary nebula, with gas puffed out by a dying star.

INSET RIGHT
**Red Rectangle**
The HST has revealed that the dying star HD 44179, commonly called the Red Rectangle because of its overall shape seen through ground-based telescopes, is not rectangular but has an X-shaped structure thought to be the result of outflows of gas and dust from the central star.

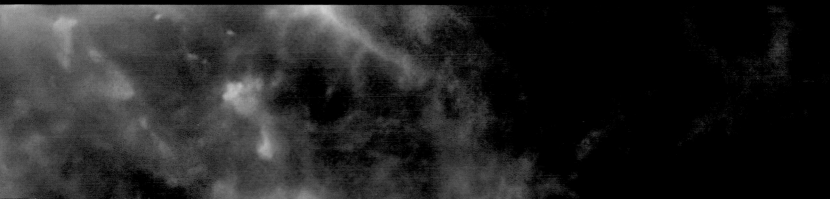

# THE BEGINNING OF THE END

Stars form out of giant clouds of interstellar gas and dust. Then, with their nuclear furnaces stoked up, they join the multitude of other stars on the main sequence. The Sun has spent around 4.6 billion years on the main sequence shining steadily, pouring out light, heat, and other radiation into space. So, will it carry on shining like it does today forever? The answer: no.

Like living things on Earth, the Sun and all the other stars have a natural life-span. Whereas the life-spans of even the oldest living things, such as California's majestic redwoods, are measured in only a few thousand years, those of stars can be measured in tens of billions of years.

The exact life-span of a star depends primarily on its mass. The more massive the star, the shorter is its life. The Sun's mass is about seven times the median mass. It is a type of star known as a yellow dwarf with a relatively long life span of about 10 billion years. This means that the Sun is now comfortably in middle age. It should carry on shining steadily like it does today for roughly another five billion years. But then, inexorably, it will begin to die.

The Sun and other similar dwarf stars die relatively quietly—we might say, with a whimper. But for stars with much greater mass it is a different story. They depart the celestial scene literally with a bang—the biggest bang in the Universe.

A star on the main sequence produces energy by transforming hydrogen into helium in its core. A star like the Sun uses around 600 million tons of hydrogen every second, but it is so massive that it has enough fuel to last for more than 10 billion years.

Eventually, though, the hydrogen in the core of the star runs out, and only helium remains. Nuclear fusion ceases. This heralds an abrupt change in the star's life. It is the beginning of the end—the star is dying.

With no radiation emanating from the star's core, gravity once again becomes the dominant force. The core begins to collapse. The potential energy released by this collapse does two things: it increases the temperature and pressure in the core, and it heats up the outer atmosphere of the star.

## COOL GIANT

The atmosphere expands greatly, making the star swell up to 30, 50, or perhaps even 100 times its original diameter. Because the star now emits radiation from a much greater surface area, its surface temperature plummets by about half, to around only 5,500 degrees Fahrenheit (3,000°C). This lower temperature makes the star produce a redder light. It becomes a red giant and leaves the main sequence behind (see page 35). When the Sun becomes a red giant in around five billion years time, it will expand 30 times or more and become 1,000 times brighter. It will probably swallow up the planet Mercury and maybe even Venus. The Earth will lose its atmosphere, and the oceans will boil away. Life will be eradicated, and our now-verdant home will become a barren cinder of a planet.

LEFT
**Red sun**
Sometimes at sunset, white sunlight turns orange and red as it passes through the lower, dustiest part of the atmosphere. In billions of years, the Sun itself will turn red as it swells up to become a giant star.

BELOW
**The generation game**
In the giant nebula NGC 3603, the HST has captured stars at different stages in their life-cycles. At upper right are dark clouds that are Bok globules, marking an early stage of star formation. In the center is a cluster of youthful hot stars. Above it is an aging blue supergiant on the brink of destroying itself in a supernova explosion.

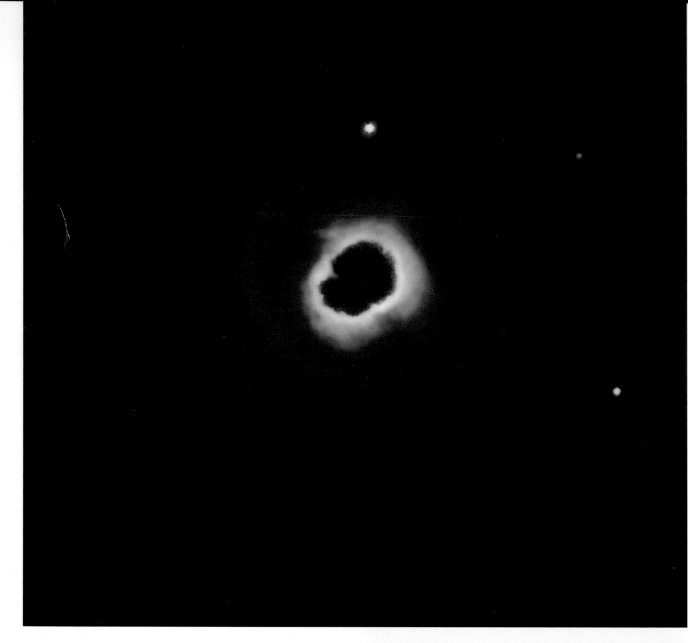

## SPOTTING GIANTS

Some of our most familiar stars are red giants, which appear in the sky with a noticeable orange–red hue. Arcturus, in the constellation Boötes (the Herdsman), is a red giant. It is the most brilliant star in the Northern Hemisphere and the fourth-brightest in all of the heavens. You can easily find it by following the curve in the handle of the Big Dipper, the most recognizable part of the constellation Ursa Major (the Great Bear). Capella in Auriga (the Charioteer) and Aldebaran in Taurus (the Bull) are two other northern giants. Aldebaran marks the baleful red eye of the bull, which appears to be charging the mighty hunter Orion in the adjacent constellation.

In the Southern Hemisphere, Gacrux, one of the four bright stars that make up Crux (the Southern Cross), is a red giant. Its warm orange color contrasts with the other three stars in the cross, which are white.

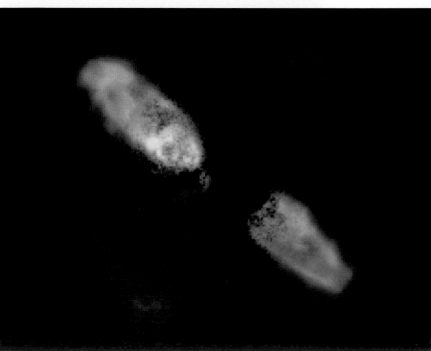

## THE HELIUM FLASH

As a red giant balloons in size, its core collapses progressively, until it is maybe only one-tenth of its original size and not much bigger than the Earth. By now, the temperature inside the core has rocketed to around 180 million degrees Fahrenheit (100,000,000°C). This is high enough to trigger new nuclear reactions, using helium as fuel. The ignition point is called the helium flash.

## TRIPLE ALPHA

The nuclei of helium atoms are also called alpha particles. In the early twentieth century, the nuclear pioneer Ernest Rutherford gave this name to particles emitted during radioactive decay, which turned out to be helium nuclei. Three helium nuclei take part in the process of nuclear fusion in a red giant's core, so it is called the triple-alpha reaction.

ABOVE
**In a telescope**
Telescopic view of the Helix Nebula, showing the region in which the HST took the close-up picture shown at right. (See page 54 for a complete view of the nebula.)

RIGHT
**Condensation**
Cometary knots line the inner edge of the Helix planetary nebula in Aquarlus. They form when hot gas ejected by the central white dwarf star condenses as it collides with colder gas ejected earlier. The head of each knot is wider across than our solar system.

In the process, first two helium nuclei fuse to form an unstable beryllium nucleus. If a third helium nucleus collides and fuses with that, it produces carbon. If a fourth collides and fuses, it produces oxygen.

Although all the hydrogen has been used up in the core, there is still hydrogen in the outer atmosphere. Temperatures are high enough immediately outside the core to cause the hydrogen to fuse. So the core takes on a kind of layered structure: carbon and oxygen accumulate in the center; around that is an inner shell in which helium is fusing; and surrounding that is an outer shell of fusing hydrogen.

## STELLAR WINDS

The heat produced in the core of a red giant rises to the surface on convection currents. These currents also carry a certain amount of core material, such as carbon, to the surface. The carbon condenses (turns into a solid) in the outermost layers, forming a soot-like dust.

As time goes by, gases in the outer atmosphere of the star gradually stream off into space as a stellar wind, just like the solar wind from the Sun. The star's pulsating—periodic expanding and contracting—sometimes exaggerates the process. The pulsations cause it to vary in brightness, becoming brighter as it contracts and heats up, dimmer as it expands and cools down.

RIGHT
**Shell of gas**
The planetary nebula NGC 2818 is in the southern constellation of Pyxis. The gas at the edge of this colorful nebula was the first ejected from the outer layers of the central dying star. Although it was expelled from all around the star and formed a sphere we see it as a two-dimensional ring.

BELOW
**The Retina**
Glowing in all colors of the rainbow, this planetary nebula (IC 4406) has a donut shape. It has been dubbed the Retina Nebula because the intricate tendrils of material we can see look like a close-up of the eye's retina.

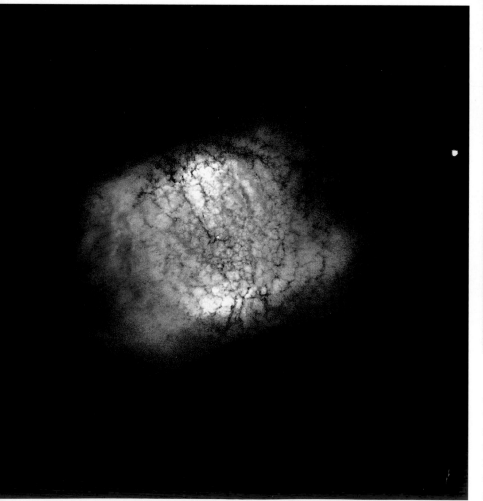

The star Omicron Ceti in Cetus (the Whale) is a classic example of a pulsating red giant. At its brightest it is easily visible to the naked eye. When at its dimmest it is beyond naked-eye detection and cannot be spotted even with binoculars. It varies from bright to dim, and back again, over a period of about 330 days. The star is usually known as Mira, which means "the wonderful." And the class of long-period variable stars it typifies are called Mira variables.

## DUSTY ERUPTIONS

Usually, the stellar wind also carries with it the sooty dust grains in the outer atmosphere. Occasionally however, the dust accumulates around the star and dims its light. Then a vigorous blast of stellar wind removes it, and the star resumes its former brightness. This also happens with supergiants. When we observe such a star, we see it as a variable with an irregular period, because the process of build-up and subsequent blowout can take weeks or even years. R Corona Borealis is a classic example. Most of the time it is bright enough to be just visible to the naked eye. But unpredictably, it can suddenly dim so that it is only visible through a telescope. It may start to recover almost immediately or do so over the course of several months.

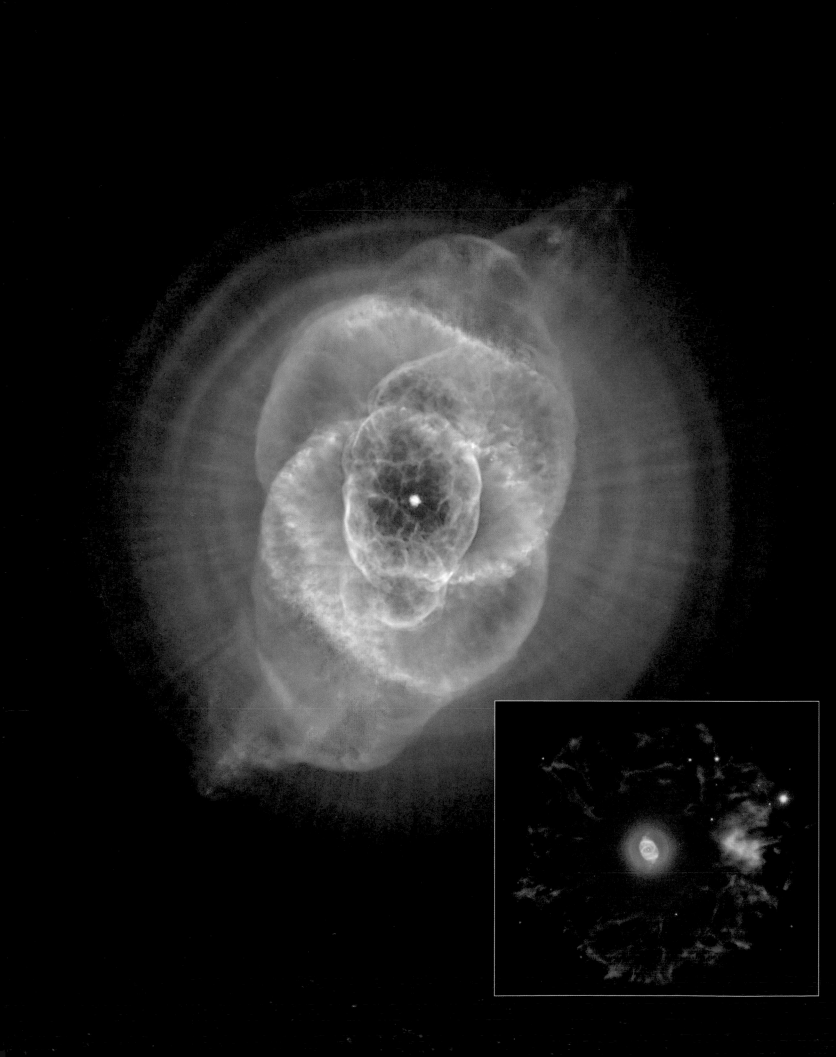

# LAST GASPS

Helium fusion in the core of a red giant lasts on average about two billion years. Then the helium runs out. Temperatures never get high enough to start fusing the carbon and oxygen that has been produced. In a star like the Sun, this marks the end of energy creation. The star is close to death.

Now come the final death throes. The super-hot core gives off a succession of shuddering death rattles that eject matter into space from the giant's outer layers. Traveling at velocities of up to 20 miles (30 km) per second, the successive waves of matter form into expanding shells. A torrent of X-rays and ultraviolet radiation streaming from the exposed core excite the ejected gas, and the shells light up with Technicolor brilliance.

## PLANETARY NEBULAE
Viewed through a telescope, these expanding shells are some of the most beautiful objects in the heavens. They are called planetary nebulae, although the term is misleading because they have nothing to do with planets. Eighteenth-century English astronomer William Herschel gave these objects their name because in the telescopes of the day they looked rather like the disks of planets.

Planetary nebulae take on a number of different forms. Some have the appearance of single or multiple spherical bubbles, like M27, the Dumbbell Nebula in the constellation Vulpecula (the Fox). We see the bubble-like shells of other planetary nebulae as rings, such as the famous Ring Nebula in Lyra (the Lyre), profiled on page 54.

## OF BUTTERFLIES AND CATS
Most nebulae, however, are quite complex in shape. Many have overlapping rings of contrasting colors, caused by successive waves of ejected material punching through earlier ejecta. Others have an hourglass shape, with the ejected gas forming lobes on either side of the dying star. The Butterfly Nebula in Ophiuchus (the Serpent-Bearer) is a beautiful example.

The Cat's-Eye Nebula in the constellation Draco (the Dragon) has an amazingly intricate structure and thoroughly deserves its name. Eleven or more concentric shells of material make a layered onion-like structure around the dying star. One explanation for this layering is that the star ejected its mass in a series of pulses at 1,500 year intervals.

With its extraordinary ability to distinguish detail, the HST has revolutionized our study of planetary nebulae. It reveals within these stellar ghosts a previously unsuspected complexity of structures and dynamic interactions.

**LEFT AND LOWER LEFT**
**The Cat's Eye**
Well named, the Cat's Eye Nebula (NGC 6369 in Draco) is one of the most beautiful and complex planetary nebulae the HST has studied. The center of the eye (main image) is surrounded by concentric rings, which are spherical bubbles of material ejected at 1,500 yearly intervals.

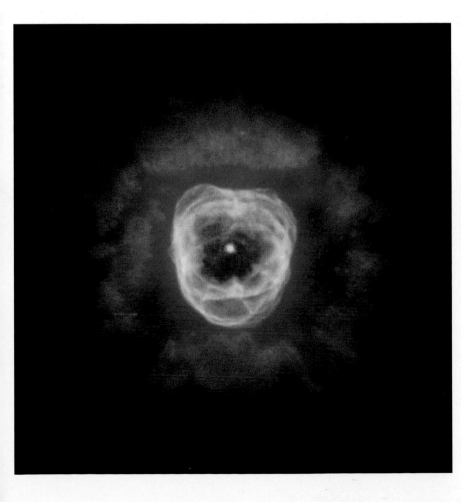

**LEFT**
**The Eskimo**
The HST's exquisite rendering of the Eskimo planetary nebula (NGC 2392 in Gemini), so called because when viewed in ground-based telescopes it resembles a face surrounded by a fur parka.

**ABOVE**
**The Egg Nebula**
Dust layers that resemble onion skins surround a dying star. Twin beams of light from the hidden star illuminate the dust. The HST image of the Egg Nebula has been artificially colored so astronomers can see how the light reflects off the smoke-sized dust particles.

**FOLLOWING PAGES**
**Southern Ring**
Planetary nebula NGC 3132, in Vela, is well named the Southern Ring. It rivals the northern Ring Nebula in Lyra in beauty. The white dwarf ejecting material is the fainter of the two stars in the center.

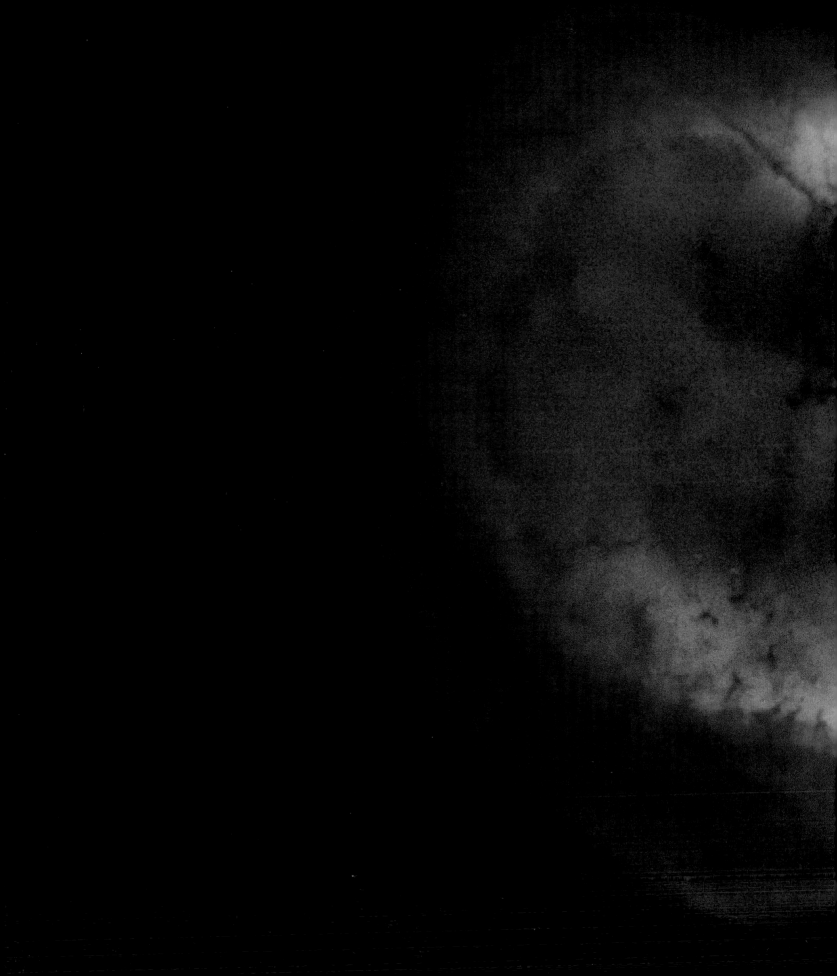

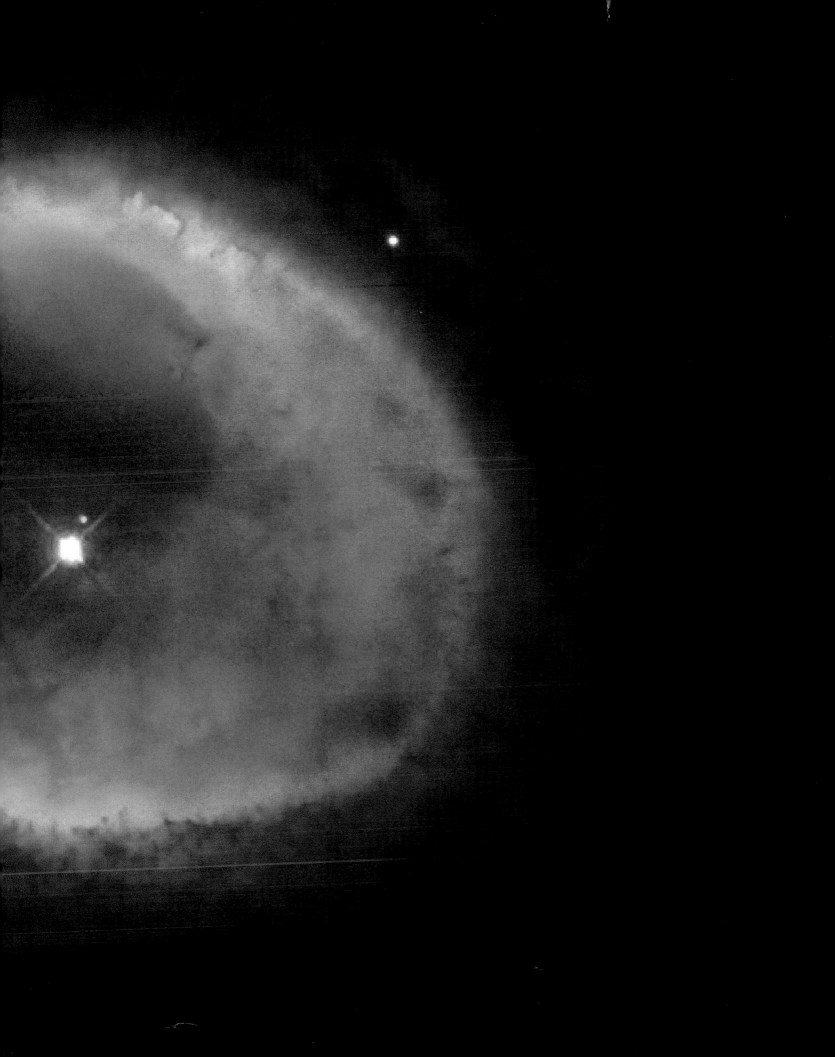

# HEAVYWEIGHT DWARFS

The outflow of matter in stellar winds and pulsations can deplete a red giant of as much as 80 percent of its mass. In tens of thousands of years, the nebula will disperse into the general interstellar medium and disappear. All that will then remain is the exposed core.

Once the core runs out of helium, no energy-producing processes can occur there. Temperatures inside are not high enough to trigger a new round of nuclear-fusion reactions with the carbon and oxygen that make up the core.

With no radiation, the core can no longer resist the inward pull of gravity, so it progressively collapses. Eventually it shrinks to about the same size as the Earth and becomes a class of star we call a white dwarf.

The gravitational energy released as the core shrinks manifests itself as heat. The temperature of the surface of the white dwarf soars to tens of thousands of degrees and beyond. The HST has spotted a white dwarf in the planetary nebula NGC 2440 with a surface temperature of 360,000 degrees Fahrenheit (200,000°C). It is one of the hottest stars we know, nearly 40 times hotter than the Sun.

## DEGENERATE MATTER

With the mass of the Sun squeezed into a volume the size of the Earth, a white dwarf is extremely dense. Just a teaspoonful of its matter would weigh tons.

Furthermore, the matter that makes up a white dwarf is not ordinary matter. Ordinary matter consists of atoms made up of nuclei, with swarms of electrons circling round them at a distance. In a white dwarf, the matter is crushed. The electrons are practically pressed up against the nuclei. The repulsion between the electrons, which are all negatively charged, prevents the body from collapsing further. This kind of matter is called electron-degenerate matter.

## CHANDRASEKHAR'S LIMIT

Typically, a white dwarf has about the same mass as the Sun. It can never have more than 1.4 times the mass of the Sun— a surprising discovery made in 1930 by Indian astronomer Subrahmanyan Chandrasekhar. If the remains of a dying star has more than 1.4 solar masses (called the Chandrasekhar limit), gravity overcomes the electron repulsion and causes further collapse (see page 62).

## THE FINAL CURTAIN

What eventually happens to a white dwarf? Most of them just quietly fade away. As they radiate away their remaining energy, they cool and dim, and their light reddens. After many billions of years they run out of energy completely. They become black dwarfs and disappear from the visible Universe. Scientists are uncertain whether any black dwarfs exist. It may be that the Universe is not yet old enough.

## THE PUP

The first white dwarf that astronomers discovered was the so-called Companion of Sirius, the Dog Star. It is nicknamed the Pup. (Astronomers call it Sirius B, and the Dog Star Sirius A.) In 1834, the German astronomer Friedrich Bessel noticed that Sirius, which is one of the nearest stars, pursues a rather erratic course through the heavens. It weaves back and forth against the background of more distant stars.

Bessel concluded that this erratic behavior must be caused by an orbiting companion star, but he still hadn't found it when he died in 1844. U.S. astronomer Alvan Clark was the first to spot the companion star 18 years later.

The companion of Sirius is beyond naked-eye visibility. Although bright enough to be visible in binoculars it gets lost in the glare of Sirius, the brightest of all the stars in Earth's sky.

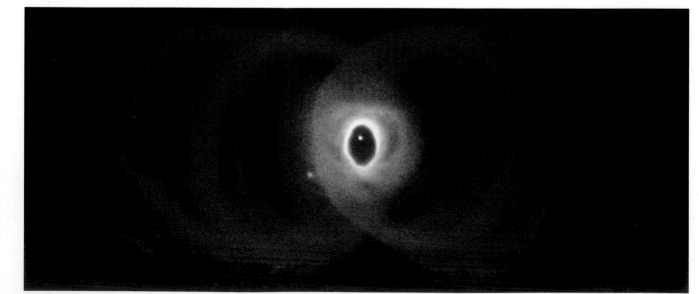

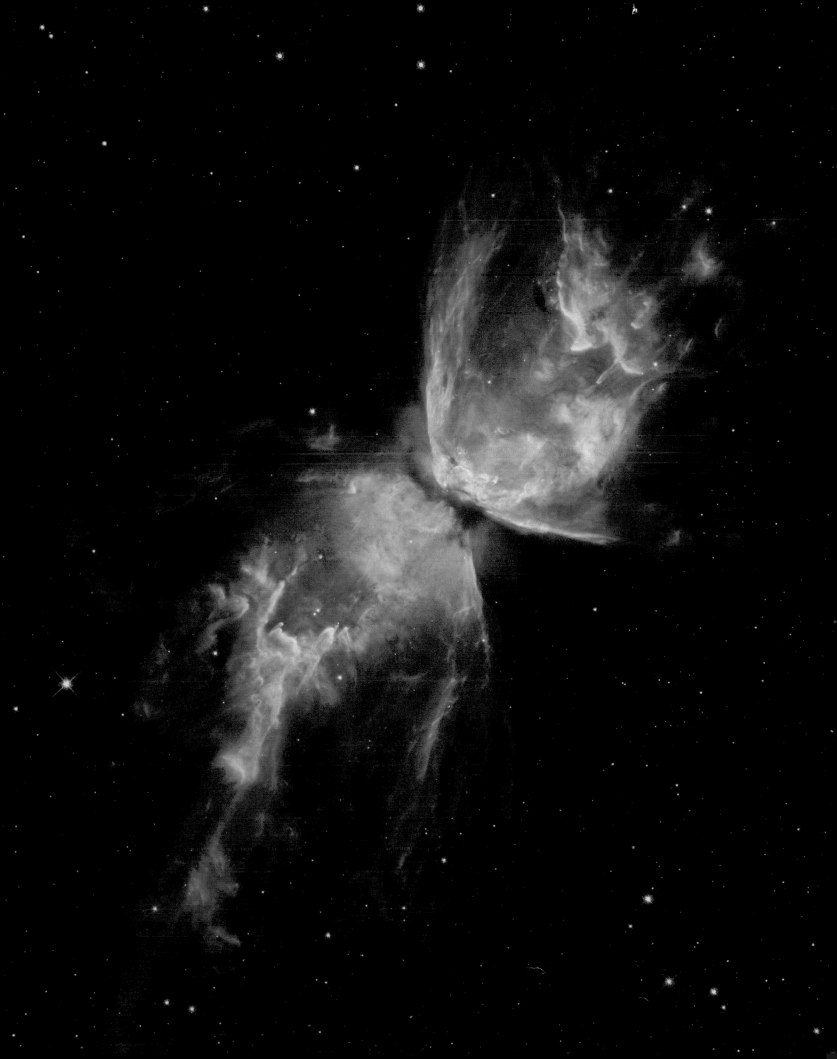

# THE RING NEBULA

Lyra (the Lyre) is one of the smallest constellations in the heavens. Greek astronomers named it for the lyre that Hermes, the messenger of the gods, made out of a tortoise shell strung with cow gut.

This tiny constellation has two outstanding features. One is its lead star Vega, the fifth-brightest star in the sky and one of the three bright stars that blaze overhead during the northern summer. The others are Deneb in the adjacent constellation Cygnus (the Swan) and Altair in Aquila (the Eagle). Together they form the famous Summer Triangle.

The other outstanding feature in Lyra is M57, the Ring Nebula. It is a planetary nebula—in fact, the definitive planetary nebula before the HST revealed the staggering beauty of so many others.

The Ring Nebula is easily located because it is sandwiched between Beta and Gamma, the second- and third-brightest stars in the constellation. It is not visible with binoculars, but even small telescopes reveal its "planetary" shape and smoke-ring appearance. Larger instruments are needed to show its central blue–white star, which is a white dwarf. It is one of the hotter types, with a surface temperature of around 180,000 degrees Fahrenheit (100,000°C).

Like other planetary nebulae, the Ring Nebula is growing as the shell of material ejected by the central star pushes farther into space. The outer edge of the ring is well defined, which makes it possible to calculate its rate of expansion as close to 12 miles (20 km) per second. Working backwards, this gives an estimated date for the origin of the nebula of around 5,500 years ago.

Several other planetary nebulae have the appearance of cosmic smoke rings, although they are not as perfect as the Ring Nebula itself. The rings are actually bubbles of expanding gas. They only look like two-dimensional rings to us on Earth because we see them at distance and they appear flattened against the plane of the sky.

RIGHT
**Captivating Ring**
Seen in exquisite detail by the HST, the Ring Nebula appears to have a cylindrical, or hourglass shape. We see it as a ring because we happen to be looking at it end-on.

BELOW LEFT
**Stellar smoke ring**
The Ring Nebula can be clearly viewed through small telescopes. Roughly 1 light-year across, it lies about 2,000 light-years away.

BELOW RIGHT
**The Helix**
This view of the Helix Nebula in (NGC 7293) is a composite of HST and ground-based images. It appears to be a simple ring shape but looks are deceiving. If viewed from the side, two gaseous disks nearly perpendicular to each other would be visible.

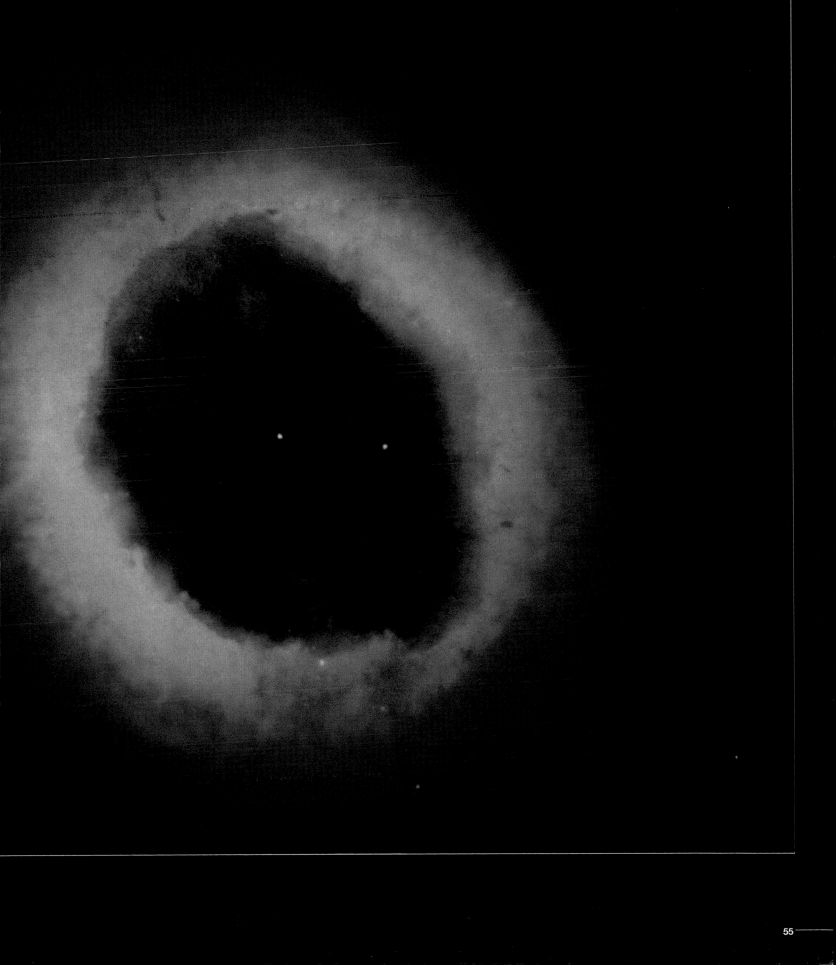

# NEW STARS

Sometimes a white dwarf forms part of a close binary, which is a two-star system, along with a larger star. It then starts to gain gas—mostly hydrogen—from its companion. Sometimes a white dwarf forms part of a binary, or two-star system, along with a larger star. It then starts to gain gas—mostly hydrogen—from its companion. As more and more hydrogen rains down, a thick, dense layer builds up, and temperatures and pressures on the white dwarf begin to soar.

Eventually, conditions are extreme enough to trigger hydrogen fusion. A massive star-wide thermonuclear explosion blasts material into space and makes the star flare up to perhaps 100,000 times its original brightness in a few days. Over the following weeks it returns to its original state, until the same process begins all over again.

Such sudden flare-ups are called novas, which means "new." Early astronomers gave this name to what seemed to be new stars appearing in the heavens, when in fact they were existing stars that had been faint but suddenly became visible.

## SUPERNOVA

Some "new stars" flare up to an even greater brilliance. They can increase millions of times in brightness and become as bright as a whole galaxy. We call these flare-up stars supernovae. On average, one supernova occurs in a galaxy about every 200 to 300 years.

The first recorded supernova seems to be the one that Chinese astronomers noted in 1054 in Taurus. We see its ghostly remains today as the Crab Nebula. The Danish astronomer Tycho Brahe spotted another supernova in our Galaxy in 1572, when it became as bright as Venus and was visible in daylight. It became known as Tycho's Star. In 1604, Tycho's protégé, the German astronomer Johannes Kepler, saw another supernova, which has been dubbed Kepler's Star.

That was the last time a supernova was seen in our Galaxy, although many have been spotted in others. Most notable was one in our galactic neighbor the Large Magellanic Cloud in 1987. Not yet operational at that time, the HST couldn't witness the event, but it has been used to study carefully the aftermath ever since.

We call the visible remains of a supernova explosion (such as the Crab Nebula) a supernova remnant (SNR). We find many other examples in the heavens, which often rival planetary nebulae in their delicate beauty.

## TYPES I AND II

There are two kinds of supernovae. A Type I supernova occurs in the same binary big-star/white-dwarf system that creates a nova. It is created when the white dwarf gains so much matter from the larger star that it can no longer support itself. It collapses and destroys itself in a supernova explosion.

A Type II supernova occurs when a star with much greater mass than the Sun dies. Type II supernovae are more common. They release much more energy than Type Is, but not in visible light. Type I supernovae appear brighter.

**LEFT**
**Stellar explosion**
These wisps of glowing gas were expelled by a star 3,000 years ago as it went through a supernova. The colossal explosion was in the nearby galaxy, the Large Magellanic Cloud. The complex structure of this supernova remnant, N132D is revealed by combining HST and images taken by the Chandra X-ray space telescope.

**RIGHT**
**Making waves**
The supernova explosion that occurs when a massive star blasts apart sends powerful shock waves rippling through the surrounding space. These waves will compress gas and make it glow, creating the luminous filaments we see in this image.

## PROFILING:

# ETA CARINAE

The HST has been used to examine closely one supermassive dying star: Eta Carinae, in the far southern constellation of Carina (the Keel). Eta Carinae is one of the biggest, hottest, and intrinsically brightest stars we know. Compared with the Sun, it is 100 times more massive, 150 times wider, and four million times more brilliant. It is also more than five times hotter than the Sun, with a surface temperature nudging 54,000 degrees Fahrenheit (30,000°C ).

Eta Carinae is just beyond naked-eye visibility but it is easily spotted with binoculars and a small telescope, and looks noticeably colored. It is one of at least a dozen brilliant stars estimated to be at least 50 to 100 times the mass of the Sun within the immense Carina Nebula.

Astronomers know that a star with such staggering statistics must be inherently unstable, and so it has proved. Eta Carinae periodically flares up and then dims, fluctuating in brightness over irregular periods of time.

English Astronomer Royal Edmond Halley (of comet fame) watched the star brighten until it was easily visible to the naked-eye in 1677. An even more spectacular flare-up began in 1835 and peaked in 1843, when Eta Carinae became the brightest star in the heavens after Sirius. Since then it has gradually dimmed.

Eta Carinae, then, is a variable star. It is called an eruptive variable because it varies in brightness when it undergoes massive eruptions from its atmosphere that blast

matter into space. It flares up because the ejected matter
is very hot and radiates abundant light.

Vast amounts of matter were ejected in Eta Carinae's
massive eruption in the 1840s. Today we see the ejecta as
an expanding red shell of nitrogen, oxygen, and some
other gases, traveling at speeds up to 2 million miles
(3,000,000 km) per hour.

The violent eruption Eta Carinae suffered is reminiscent
of a supernova explosion, but it was only a hint of what is
to come. When this doomed star really goes supernova
and blows itself to bits—which could happen any time—it
will be a truly amazing sight.

RIGHT
**Flare-up**
**Visible in this HST
image of Eta Carinae is
a double-lobed cloud
of gas and dust. This is
the material that
erupted from the
hidden central star
about 160 years ago.
The star then flared up
to become one of the
brightest yet found in
the heavens. In fact, it
is now believed that
there are two (rather
than one) central stars.**

# THE DEATH OF MASSIVE STARS

Massive stars destroy themselves in a Type II supernova explosion. As was noted earlier, a star with about the same mass as the Sun rides the main sequence for billions of years, swells up into a red giant, puffs off its outer layers and then shrinks to become a tiny white dwarf.

But stars much more massive than the Sun live life in the fast lane. They consume their hydrogen fuel voraciously and spend only a few tens of millions of years on the main sequence. The biggest stars of all may spend less than a million years there.

As with a star like the Sun, a supermassive star stays on the main sequence while hydrogen fuses into helium in its core. When the hydrogen runs out, it starts swelling up, becoming cooler and redder. Because it is so massive, it swells up beyond the size of a red giant to become a supergiant, with hundreds of times the diameter of the Sun.

## INSIDE A SUPERGIANT

Like in a red giant, the core of a supergiant starts collapsing when it runs out of hydrogen. The collapse raises the temperature and initiates the fusion reactions that turn the helium into carbon and oxygen.

Once all the helium has been used up, the core begins to collapse again. Because the core in a supergiant is so massive, its collapse releases prodigious amounts of energy, pushing temperatures up to hundreds of millions of degrees. This is high enough to make the carbon and oxygen fuse to produce heavier elements, such as neon, magnesium, silicon, and sulfur. These elements fuse in turn. The final product is iron. Elements heavier than iron are formed in supernovae.

The rapidity with which these elements fuse is staggering. The carbon "burn" may take only about a thousand years, the oxygen burn just a year and the final silicon burn to form iron only a few days.

## THE MOTHER OF ALL EXPLOSIONS

Once the entire core of a supergiant has been converted into iron, its enormous mass causes it to collapse. The collapse is so rapid that the rest of the star crashes down on itself as well. The release of energy is catastrophic and blasts the star to smithereens. It becomes a supernova.

The temperatures and pressures produced in this cataclysmic event convert iron into a succession of heavier elements. In fact, all the elements heavier than iron that exist in nature were forged in supernovae. These mothers of all explosions blast newly made elements into interstellar space, where they make their way into the vast nebulae that spawn new stars.

Without the production of heavier elements within the stars, and the celestial re-cycling whereby the spectacular deaths of massive stars seed the Universe with material for new stars, humans would not exist.

## SUPERNOVA SN 1987A

On February 23, 1987, astronomers spotted a bright "new star" in southern skies. It was a brilliant supernova, designated 1987a. Over the next weeks it was easily visible to the naked eye. It was the brightest supernova seen since Kepler's Star of 1604. But SN1987a was not in our Galaxy; SN 1987a was in a neighboring galaxy, the Large Magellanic Cloud, some 160,000 light-years away.

The star that went supernova was identified as Sanduleak −69°202. It was a blue supergiant estimated to have had a mass 20 times that of the Sun. Shortly after the HST went into orbit in 1990, it spied a ring of matter that had been blasted into space by SN 1987a. It has been charting the incredible convolutions of this expanding ring ever since.

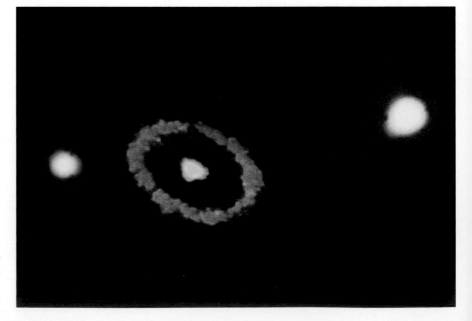

## END OF THE LINE

Exactly what happens to the collapsing core of a massive star in the aftermath of a supernova explosion depends on its mass. It will either become a neutron star or that most awesome of celestial objects—a black hole.

If the core is between 1.4 and three times the mass of the Sun, it becomes a neutron star. The force of collapse creates a new kind of matter. It forces the electrons in atoms into the nucleus, where they merge with the protons and turn into neutrons. The core becomes a mass of neutrons, tightly compressed together—a form of matter called baryon degenerate matter. (Neutrons belong to a class of subatomic particles called baryons.)

Such a neutron star is tiny, averaging only about 12 miles (20 km) across. Yet it contains between 1.4 and three solar masses. So it has the most incredibly high density—millions of times greater even than a white dwarf.

## PULSATING SIGNALS

The small, dense neutron star that remains after a supernova explosion spins furiously. Its powerful magnetic field spins as well, and funnels the star's energy into powerful beams of radiation that emanate from the two magnetic poles.

As the star rotates, these beams sweep around in space, rather like those from a lighthouse. If the beams are in our line of sight, we detect pulses of energy every time they sweep past. We call these pulsating bodies pulsars.

## LITTLE GREEN MEN

Working at Britain's Cambridge Radio Astronomy Observatory, astronomer Jocelyn Bell-Burnell discovered the first pulsar in 1967 by its radio waves. Never before had such a regularly pulsating radio source been detected. Initially baffled over what it could be, the Cambridge astronomers dubbed it LGM (for Little Green Men), likening it to the kind of signals an alien civilization might beam into space.

Since then, more than 1,800 pulsars have been discovered. Most give off energy as radio waves, but some give off energy in other forms, including visible light, X-rays, and gamma rays.

The fastest pulsar, PSR J1748-2446ad, was discovered in January 2006. It is in a globular cluster of stars called Terzan 5, in the constellation Sagittarius and is spinning 716 times per second. It whirls round faster than the blades of a kitchen blender. Try to imagine something the size of New York spinning around that fast!

## BLACK HOLES

After a supernova, if the collapsing core has a mass greater than about three solar masses, it becomes a black hole. When such a massive core collapses, the powerful gravitational forces compress even neutrons, causing the body to go beyond the neutron-star stage. As the core gets ever smaller, the surface gravity gets ever larger, and the velocity needed for anything to escape from it escalates.

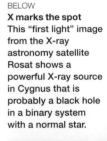

Eventually the surface gravity of the collapsed core becomes so high that not even light can escape from it. At this point it becomes a black hole and disappears from the observable Universe. The circle that marks the edge of a black hole—the boundary between it and the visible Universe—is called the event horizon. Its radius, called the Schwarzschild radius, depends on the mass of the collapsing body. Usually it measures just a few miles.

What lies within the event horizon is, of course, impossible to fathom. The theory is that the collapsing core continues to shrink until it becomes an infinitely small point of infinite density known as a singularity. The reality could be different—we will probably never know.

## X MARKS THE SPOT

Looking for black holes in the blackness of space is obviously impossible, but there are ways to detect them. When a black hole exists in a binary, two-star system, it attracts gas from the other star with its powerful gravity. As the two bodies orbit each other, the gas first swirls round the black hole, forming what is called an accretion disk.

The material in the disk swirls round very fast, and becomes searing hot because of friction between the particles. It gives off energy as X-rays before spiraling into the black hole. And we are able to detect these X-rays.

Cygnus X-1, a powerful X-ray source in the constellation Cygnus, is believed to come from such a binary black-hole system. It was the first object that astronomers suspected was a black hole, and they estimate that it contains the mass of nearly 20 Suns.

There are some black holes that have a mass that is millions of times the mass of the Sun. These supermassive black holes are found in the centers of galaxies, where they were produced by the collapse of enormous gas clouds, rather than by dying stars. They are the extremely powerful "engines" that are responsible for the extraordinary energy output of quasars and other active galaxies (see page 82).

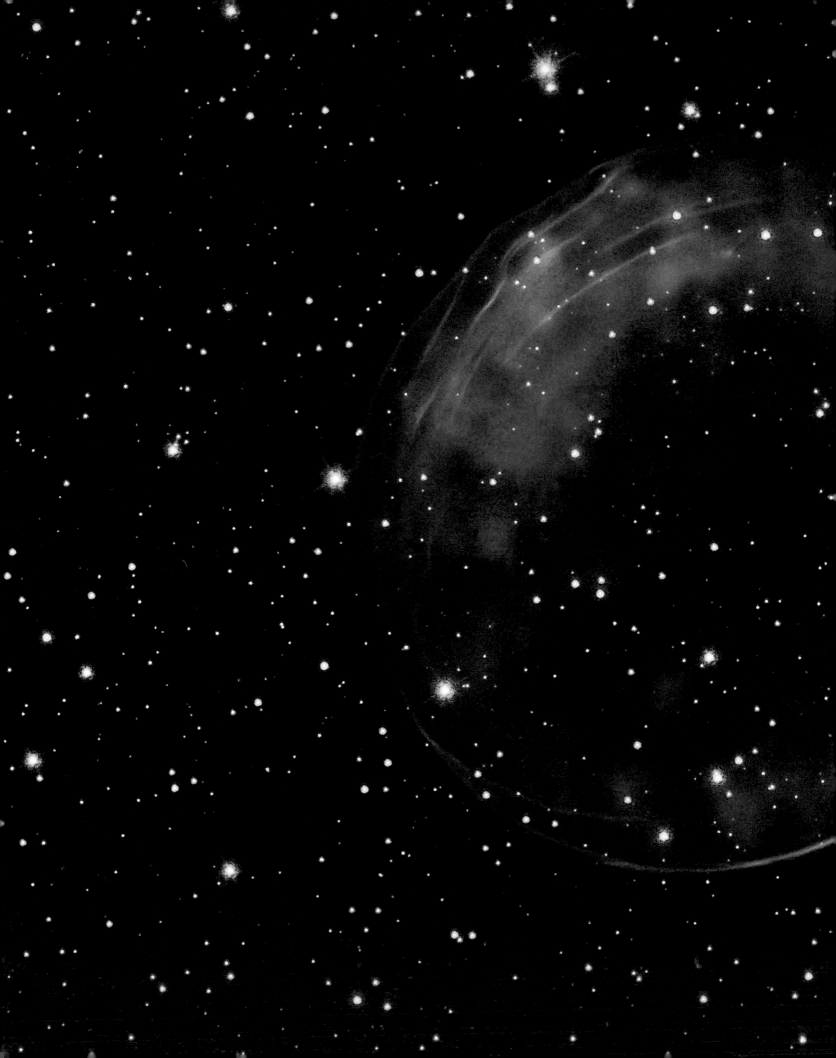

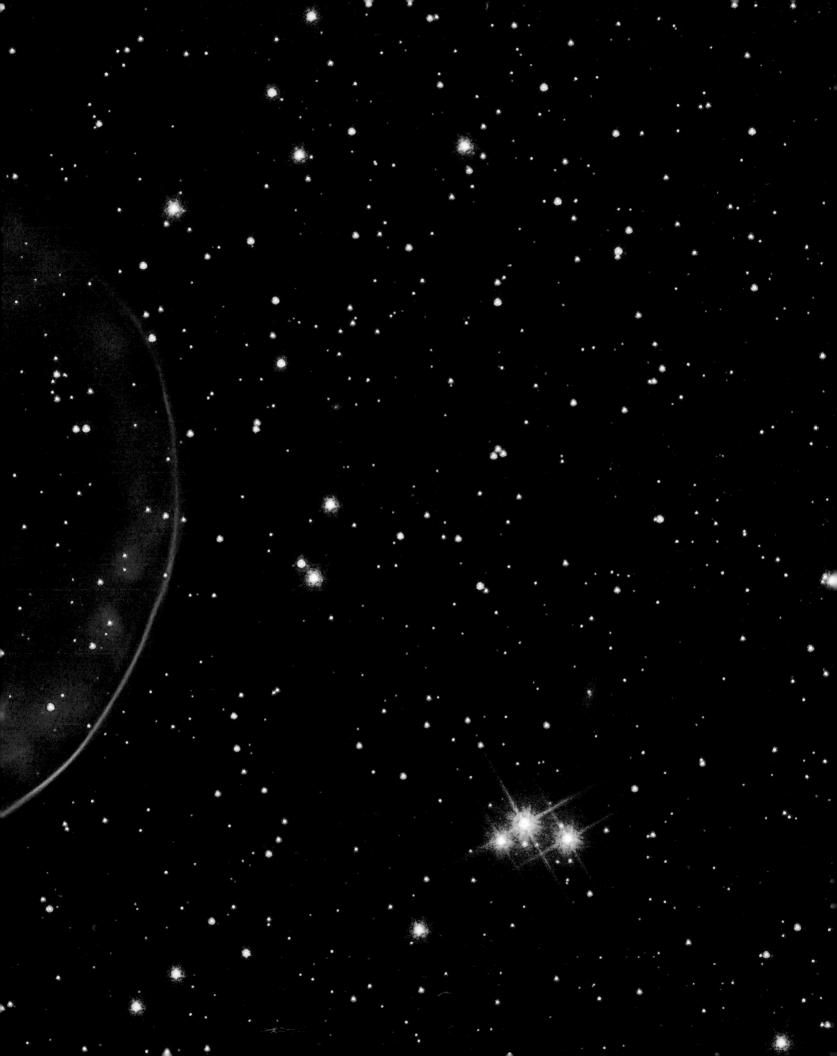

# 3 | Gregarious Galaxies
**The HST reveals the beauty of these great star islands in space**

ABOVE
**Starburst galaxy**
Young stars are being born ten times faster in the central region of the starburst galaxy M82 than inside the Milky Way. Superwinds from hot, bright stars help form towering plumes of hot gas, seen here in red, above and below the disk of the galaxy.

INSET LEFT
**Barred spiral**
The HST's sharp eye has revealed myriad fine details in barred spiral galaxy NGC 1300. Blue and red supergiants are seen in the spiral arms, and dust lanes trace out fine structures in the disk and central bar. Numerous more distant galaxies are seen in the background.

INSET RIGHT
**Edge-on**
The galaxy NGC 5866 is nearly edge-on to our view. Face-on it is a flat disk with a spiral structure. Edge-on we see a bright nucleus split by a dust lane. A blue disk of stars runs parallel to the dust lane, and a transparent outer halo dotted with globular clusters surrounds it all.

# THE MILKY WAY

In whichever direction you look in the night sky, you see twinkling stars scattered in the velvety blackness of space. So is this what the Universe is like—stars scattered haphazardly in space? Indeed it is not. If you could take a ride on an interstellar spaceship, head for the stars and then keep going, you would eventually leave the stars behind. Looking back, you would see that all the stars in the sky are grouped together, forming a great star island in a vast ocean of space. And looking in other directions, you would see other star islands, with empty space in between. We call these separate star systems galaxies. Edwin Hubble was the first to prove the existence of external galaxies beyond our own and study them in detail.

Our own galaxy is called the Milky Way, but we often also refer to it simply as "the Galaxy." Like most other galaxies, it is a collection of billions of stars, bound together by gravity. It has a disk-shape with spiral arms and rotates slowly. From a distance it would look like a spinning wheel-shaped firework. Many galaxies are similar to it, but others are different. The most extraordinary ones pump out unbelievable amounts of energy. They include quasars—some of the most amazing and remote objects in the Universe.

In the study of galaxies, the HST reigns supreme. With its incomparable clarity of vision and sensitivity, it provides panoramic vistas of deep space showing galaxies by the thousands. It can also zero in on individual star systems to tease out the most minute details.

Look at the heavens on a clear, moonless night, and you'll see a faint, misty band of light arching across the sky. In the Northern Hemisphere, it runs through the constellations Cassiopeia, Perseus, Cygnus, and Aquila, and in the Southern Hemisphere, Crux, Centaurus, Scorpius, and Sagittarius.

What is this tenuous misty band? The ancient Greeks said it was a stream of milk spurting from the breast of the goddess Hera, the long-suffering wife of Zeus, the philandering king of the gods. They called it Kiklos Galaxias, meaning the milky circle. We know it today as the Milky Way. It was the Italian Galileo who first discovered the nature of the Milky Way, when he turned his telescope on the heavens in the winter of 1609–10. He saw that it was made up of countless numbers of faint stars, seemingly packed tightly together. If you look at the Milky Way through powerful binoculars, you'll see that he was right.

Later the following century, astronomers such as William Herschel began to realize what the Milky Way really is. It is a view from inside a layer of stars that form our star system—our Galaxy. By counting stars in regions of the sky on either side of the Milky Way, Herschel concluded that the Galaxy was lens-shaped: thickest at the center and thinnest at the edge. He was not far wrong.

## OUR GALACTIC HOME

Our Galaxy, also called appropriately the Milky Way, is shaped like a disk with a huge bulge of stars in the middle. This bulge, also known as the nucleus, is bar-shaped and has arms of stars spiralling out from its two ends. For this reason the galaxy is classed as a barred-spiral. Galaxies with a round bulge and arms are spirals (see page 74).

The size of the Galaxy is astonishing. It measures 100,000 light-years from edge to edge. The nucleus is about 6,000 light-years thick, while the disk averages only about a third of this.

The number of stars it contains is uncertain; it is thought to have at least 500 billion stars. The Sun is located on one of the spiral arms, about 25,000 light-years from the center. The center lies in the direction of the constellation Sagittarius. And this is where the Milky Way appears brightest in the sky.

RIGHT
**Face-on**
Fifty-one individual HST exposures have been combined to produce this stunning image of the gigantic Pinwheel Galaxy (M33 in Triangulum). It is twice the diameter of the Milky Way Galaxy and contains at least one trillion stars, but gives an idea of what we'd see if we could view our galaxy from afar.

FOLLOWING PAGES
**Heart of the Milky Way**
Observations made by three space telescopes—Hubble, Spitzer, and Chandra—were combined to produce this image of the central Milky Way. The bright white region (center right) marks the galaxy's heart. Yellow regions are stars being born; red are Spitzer's infrared view of dust clouds; blue highlights particularly hot regions located by Chandra's X-ray eyes.

RIGHT
**Night sky view**
A dazzling path of closely packed stars, bright clusters, and colorful nebulae crosses the night sky. Called the Milky Way, it is our view into the Galaxy's disk and along its plane. Here we look towards the center of the Galaxy.

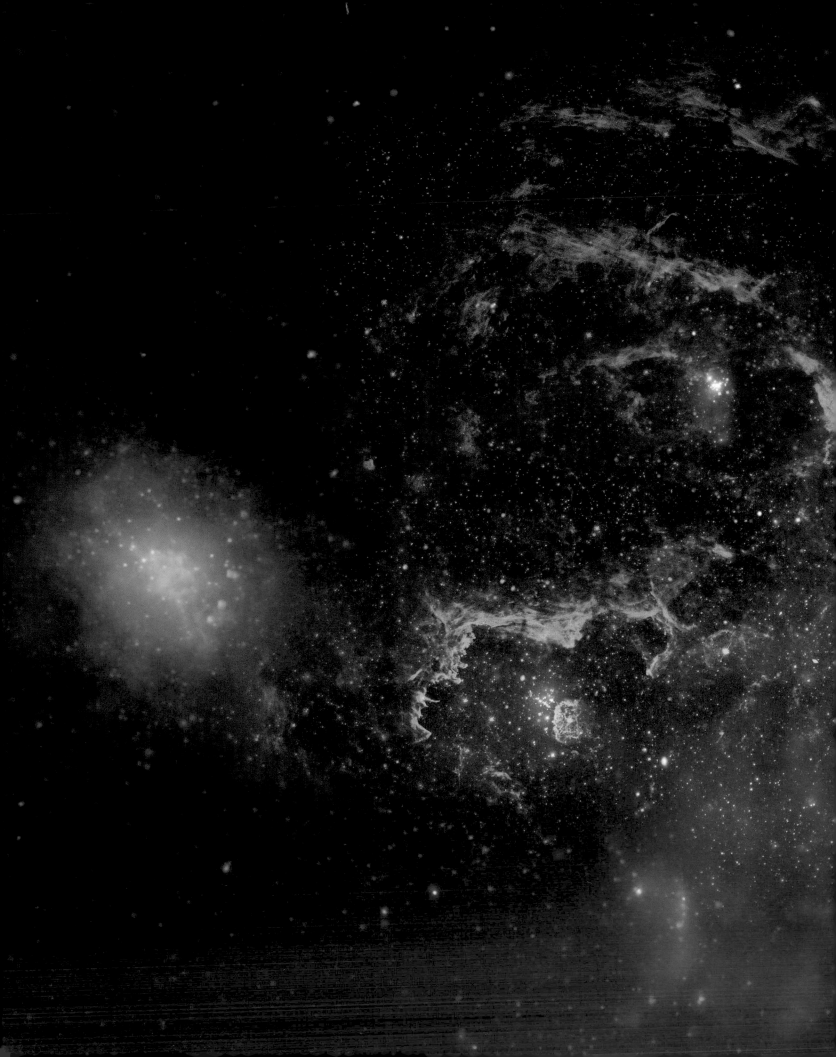

## IN THE CENTER

The central bulge of the Galaxy contains mostly old red and yellow stars. There is comparatively little interstellar matter, and star formation is limited.

Until the last decades of the 20th century, what lay in the interior of the bulge was a mystery, because thick clouds of gas and dust obscured our view. Radio and infrared studies of the region have now revealed a host of fascinating features.

At the exact center is a powerful radio source known as Sagittarius A*. It appears to mark the site of a black hole with the mass of about 3.5 million Suns. Astronomers believe that supermassive black holes like this lurk at the center of most galaxies. Magnetism is also a dominant force in the galactic center, creating strange structures such as the Arc.

## THE SPIRAL ARMS

Using radio telescopes to penetrate the obscuring clouds in the Milky Way, astronomers have created a detailed map of the Galaxy. These radio surveys have revealed the curved, spiral "arms" that make up the disk. Young, bright stars and bright and dark nebulae make up the shape of the arms.

The galaxy's two major arms, the Scutum-Centaurus and the Perseus, start from either end of the central bar of stars. Two lesser arms, Norma and Sagittarius, lie between the two major ones. The Sun is located in a partial arm called the Orion Arm and sometimes the Orion Spur. It is between the Sagittarius and Perseus arms. Our neighborhood is host to stars of some of the most familiar constellations, such as Orion, Taurus, and Cygnus. Clusters such as the Hyades and Pleiades lie relatively close, as do some of our most familiar nebulae, such as the Orion, Helix, Dumbbell, and North America.

Farther away from the galactic center the Perseus Arm contains several well-known supernova remnants, including the Crab Nebula and Cassiopeia A. The Rosette Nebula in Monoceros and the double star cluster in Perseus are among other delights.

FAR LEFT
**On the radio**
This radio image of the center of the Galaxy shows the curved feature called the Arc, which seems to be part of a ring of highly magnetized gas (the radio lobe). The white point is the radio source Sagittarius A*.

LEFT
**On the Orion Arm**
The Dumbbell planetary nebula is one of the many familiar night-sky objects in our local celestial neighborhood. It lies about 1,000 light-years away. Its characteristic dumbbell shape is more evident in less powerful telescopes than the HST.

ABOVE
**Quintuplets**
Near the center of the Galaxy, the HST has spotted a group of giant stars, called the Quintuplet Cluster. Ten times larger than typical open clusters in the spiral arms, it is home to one of the most luminous stars that we know, the Pistol Star (see page 199).

But it is on the Sagittarius Arm, closer to the galactic center than the Orion Arm, that we find some of the most spectacular night-sky objects. Great billowing clouds of interstellar matter mark regions of intensive star birth in the part of the arms closest to us. They include the Eagle, Omega, Trifid, and Lagoon Nebulae. The unstable giant star Eta Carinae threatens to go supernova, while the sparkling stars in Crux's Jewel Box cluster could not be more appropriately named.

## ON THE OUTSIDE

Most of the matter in the Galaxy resides in the stars in the bulge and spiral arms, and in the interstellar medium—the gas and dust between the stars. But about one hundred and fifty globular clusters circle outside the bulge and disk, and there are also large amounts of invisible dark matter within a great spherical halo that envelops the entire Galaxy.

# CLASSIFYING THE GALAXIES

Our home Galaxy—a rotating disk of hundreds of billions of stars that measures thousands of light-years across—is one of billions of such galaxies in the Universe. It is a barred spiral galaxy because arms of stars spiral out from a bar-shaped region in the center of the disk. Similar galaxies but with a round-shaped center are spirals. Other galaxies consist only of a spherical or elliptical bulge of stars and lack the curved arms of spirals. They are known as ellipticals.

Galaxies with no distinctive shape or form are classed as irregulars. Edwin Hubble devised the method of classifying regular galaxies according to their shape in his so-called tuning fork diagram.

## ELLIPTICAL GALAXIES

More than half of all galaxies are ellipticals. They consist mainly of old stars and contain little free gas or dust, and so little or no star formation takes place within them. Elliptical galaxies vary widely in size and mass. The largest and some of the smallest galaxies are ellipticals. The giant elliptical galaxies at the heart of galaxy clusters measure many hundreds of thousands of light-years across.

At the other end of the scale are dwarf elliptical galaxies just a few hundred light-years across. Many of our galactic neighbors are dwarf ellipticals, some containing only a few hundred thousand stars, rather than billions.

In the Hubble classification, ellipticals are denoted E, followed by a number from one to seven describing how round or oval they are. E0 is a near-spherical galaxy and E7 the most flattened oval.

## SPIRAL GALAXIES

Spirals and barred spirals have essentially the same structure and composition; a bulge at the center, and stars and dust carried on arms spiraling out. The bulge contains mainly older stars, with little gas and dust, and is orbited by up to a few hundred globular clusters. The arms, by contrast, are populated

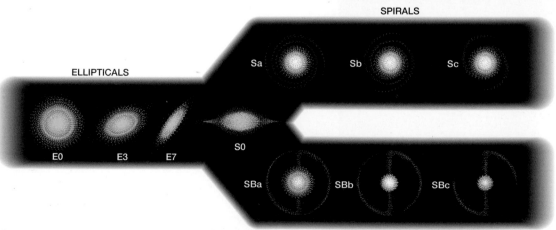

SPIRALS

Sa  Sb  Sc

ELLIPTICALS

E0  E3  E7  S0

SBa  SBb  SBc

BARRED SPIRALS

LEFT
**Hubble's tuning fork**
Edwin Hubble's system of classifying regular galaxies into ellipticals, spirals, and barred spirals.

ABOVE
**Dusty spiral**
A beautiful spiral galaxy (NGC 4414) in Coma Berenices, which has copious amounts of dust in its spiral arms. As is usual in spirals, the central bulge has older, yellower stars, while the arms have many more younger, bluer ones.

ABOVE
**Bar of stars**
The nearby galaxy
NGC 1672 is a
prototypical barred
spiral galaxy. Its
spiral arms are
attached to the
two ends of a
straight central
bar of stars. At
either end of the
bar are regions
of concentrated
star formation.

ABOVE RIGHT
**Irregular galaxy**
Galaxy NGC 1569
sparkles with the light
from millions of newly
formed stars. They are
being produced at a
rate 100 times the
star-formation rate of
the Milky Way. And
this production rate
has been almost
continuous for the last
100 million years.

RIGHT
**Elliptical galaxy**
There are several
trillion stars within
the huge elliptical
galaxy M87. They are
predominantly older,
yellow stars. A jet of
material ejected from
the galaxy's core by
the power of a
supermassive black
hole is visible at
center right.

mainly by young stars, and star formation is rampant among
the turbulent gas clouds of the interstellar medium.

Spirals do not vary as much in size as ellipticals. The Milky
Way seems to be of about average size. In our galactic
neighborhood, the Triangulum Galaxy is half its size, while the
Andromeda Galaxy (see page 112) is half as big again.

In the Hubble classification, spirals are denoted S and
barred spirals SB. The letter a, b, or c follows, indicating the
openness of the spiral arms: an Sa galaxy has the most closed
arms, an Sc the most open. Our own Galaxy was once
thought to be a spiral but in recent years it has been re-classed
as an SBc; a barred spiral with open arms.

## IRREGULAR GALAXIES

Irregulars, with no particular shape, are not all that common. In
general, they are comparatively small and contain relatively few
stars. Yet they are rich in clouds of gas and dust, in which stars
are being born. The two Magellanic Clouds visible to the naked
eye in far southern skies are irregulars (see page 80).

**Colorful spiral**
The spiral galaxy NGC 3982 is about a third the size of our Milky Way. Its bright nucleus contains older stars; whereas, the winding arms in its disk house pink star-forming regions of glowing hydrogen, newborn blue star clusters, and dark lanes of dust.

**Edge-on**
If we could view our Galaxy edge-on from far out in space, it would look like this. This galaxy is NGC 4013 in Ursa Major. The dark band that cuts it in two is thick dust blotting out any light given off from background stars.

**Old and young**
Elliptical galaxy NGC 4150 looks serene from a distance (left) but when its core is seen close-up (above) it reveals recent activity. Most of the galaxy's stars are about 10 billion years old but the core's dark strands of dust and blue regions of recent starbirth tell a different story. The stars here are less than a billion years old and they probably formed during an encounter with a smaller galaxy.

**Newborn stars**
The Hubble's Wide Field Camera 3 took this detailed view of the galaxy M83 in August 2009. It reveals starbirth in one of the galaxy's spiral arms. The newborn stars are largely in clusters on the edges of dark dust lanes and are identified by reddish glowing hydrogen gas. The bright whitish region at far right is the galaxy's core.

# PROFILING:
# MAGELLAN'S CLOUDS

The early sixteenth century was an exciting time to be alive if you were of an adventurous disposition. Italian-born Christopher Columbus had recently sailed from Spain across the Atlantic and discovered a New World in 1492. Vasco da Gama had sailed from Portugal round the Cape of Good Hope to India in 1498. This voyage fired the imagination of his young fellow countryman Ferdinand Magellan, then just 18 years old.

Twenty-one years later, now an experienced seaman and navigator, Magellan commanded an expedition that would become the first to circumnavigate the globe. Alas, he never completed the voyage because he was killed in a skirmish with tribesmen in the Philippines in 1521.

Navigating for months under southern skies, Magellan would have become familiar with the dazzling southern constellations, particularly Crux (the Southern Cross). He would also have noticed the two misty patches visible nearby, and these were since named after him. They are the Large and Small Clouds of Magellan, or Magellanic Clouds.

Also named Nubecula Major and Minor, the Magellanic Clouds look like nebulae, but aren't. They are neighboring galaxies—two of only three galaxies we can see with the naked eye. The other is Andromeda (see page 112).

The Large Magellanic Cloud (LMC) is closest to us, at a distance of about 160,000 light-years. It contains much the same mix of stars and gas as our own Galaxy and has some structure. But at only about 30,000 light-years across, it is not big enough to develop into a spiral galaxy such as our own. One striking feature of the LMC is the Tarantula Nebula, named for its spidery appearance. It is one of the biggest and brightest nebulae we know and is easily visible through binoculars. The Small Magellanic Cloud (SMC) is only two-thirds as big across as the LMC and is about 30,000 light-years farther away.

The Magellanic Clouds are not only close neighbors in space, they are actually companions—satellites—of our own Galaxy. They circle our Galaxy once every 1.5 billion years or so, traveling in an elliptical orbit. Every time they make their closest approach, our Galaxy's powerful gravity attracts some of their stars and gas. As a result, the SMC is already showing signs of breaking apart. In time, both galaxies will be absorbed by our own.

RIGHT AND FAR RIGHT INSET
**Heart of the Tarantula**
**The Tarantula Nebula, also known as**
**30 Doradus, is the largest stellar nursery**
**in the local Universe. Fifteen HST images**
**have been combined to reveal the center**
**of this hotbed of star formation.**

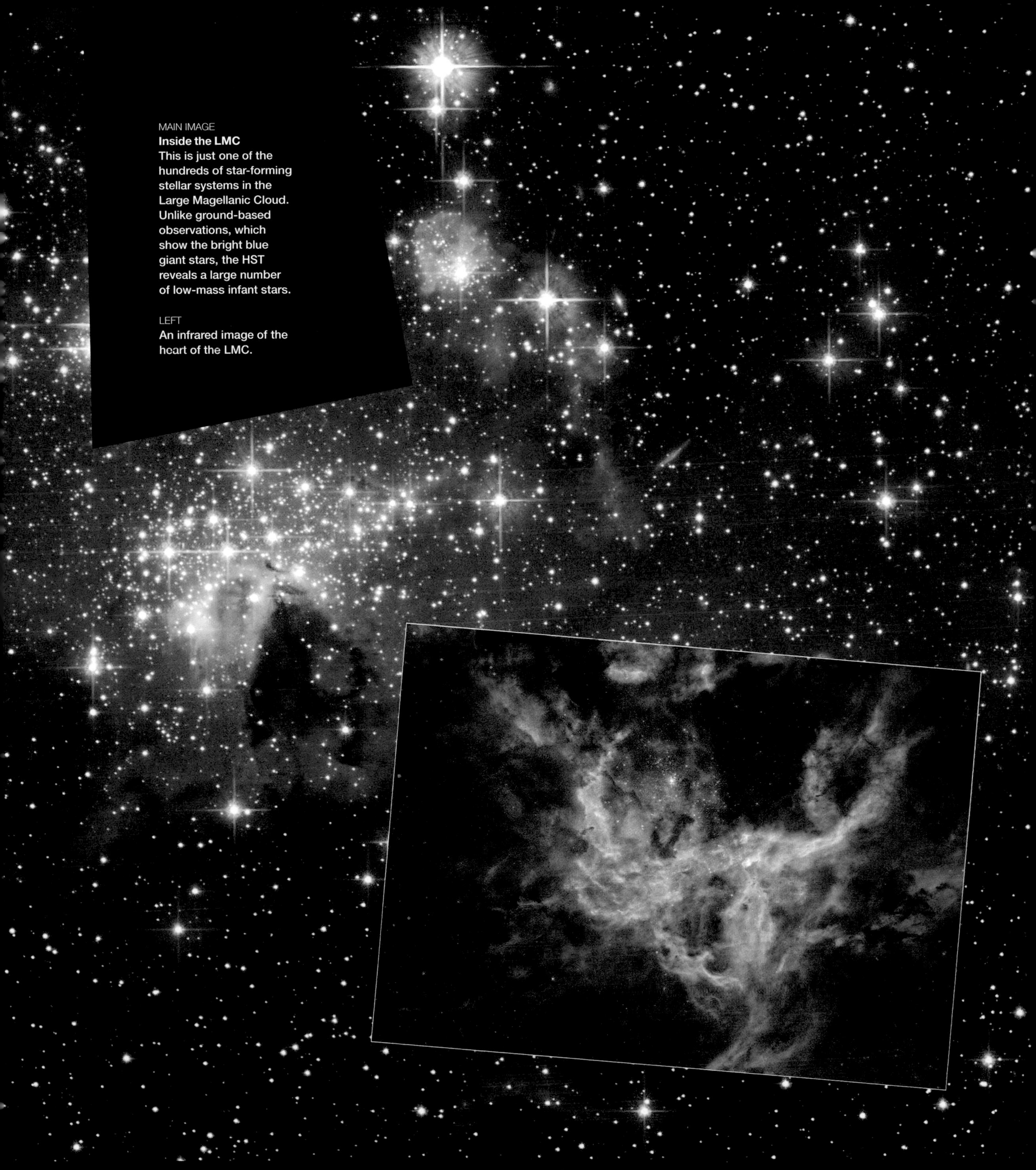

**Inside the LMC**
This is just one of the
hundreds of star-forming
stellar systems in the
Large Magellanic Cloud.
Unlike ground-based
observations, which
show the bright blue
giant stars, the HST
reveals a large number
of low-mass infant stars.

LEFT
An infrared image of the
heart of the LMC.

# ACTIVE GALAXIES

Galaxies give off energy as light, heat, and other invisible radiation, such as X-rays and radio waves. Most of them give out the energy you would expect from a collection of billions of stars. But just a few—about one in 10—give off exceptional energy, maybe millions of times greater than you might expect. They seem to pump out this energy from a tiny region at their center not much bigger than the solar system. Astronomers call these galactic mavericks active galaxies.

In 1943, U.S. astronomer Carl Seyfert noted that some spiral galaxies have exceptionally bright centers. We now know that these Seyfert galaxies are one kind of active galaxy. Since then, several other kinds of active galaxies have been discovered, including radio galaxies, quasars, and blazars.

## SEYFERT GALAXIES

These active galaxies are among the most obvious because they mainly emit their exceptional energy at visible wavelengths. Analysis of their spectrum indicates that they contain clouds of hydrogen gas swirling at very high speeds around the galactic center.

## RADIO GALAXIES

At radio wavelengths, these active galaxies appear among the biggest objects in the sky. The radiation seems to come not from the galaxy itself but from regions—radio lobes—on either

ABOVE
**Swan song**
The second most powerful radio source in the heavens is located in Cygnus (the Swan). Called Cygnus A, it is an active galaxy that pumps out a million times more energy at radio wavelengths than our own Galaxy. The energy is broadcast mainly from two lobes, each 200,000 light-years from the galaxy's center.

BELOW
**Along the jets**
Combined HST and radio images of the radio galaxy 3C-368 show a string of bright knots that may be stars or pockets of dust. This suggests that the jets streaming from black holes at the center of active galaxies might trigger star formation along their path.

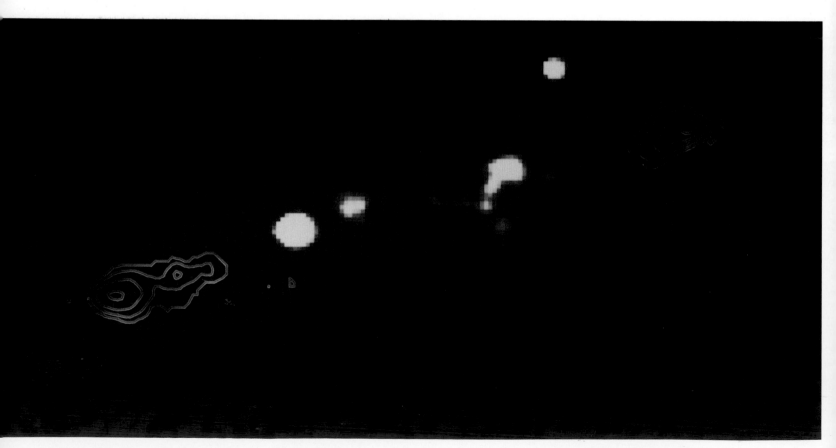

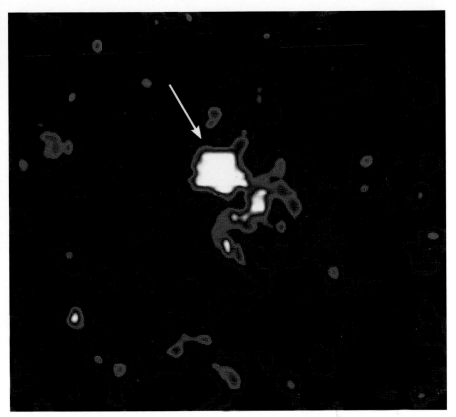

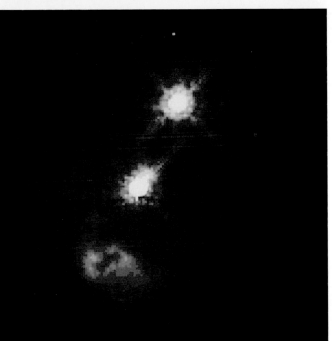

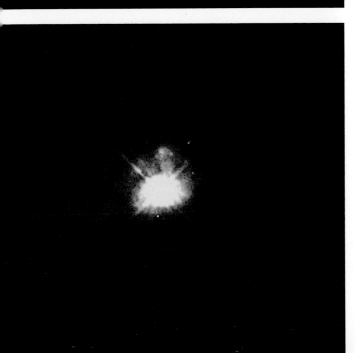

side and very far out. Although the galaxy may measure only 100,000 light-years across, from lobe to lobe the radio source may span millions of light-years.

## QUASARS

In the third Cambridge catalogue of radio sources is one in the constellation Virgo, numbered 273, so it is identified as 3C-273. In 1963, when the Moon passed in front of this source and blotted out its signals, astronomers identified it with a star and took its spectrum.

When U.S. astronomer Maarten Schmidt checked the spectrum, he found that it was like that of no other star he had ever seen. This "star" had a huge red shift of spectral lines that placed it at a distance of more than two billion light-years!

3C-273 was clearly no ordinary star, but quite a different body altogether. To be visible at such a distance, it had to be hundreds of times brighter than ordinary galaxies like the Milky Way. Because it appeared star-like, this newly identified body was named a quasi-stellar radio source, or quasar.

Since that time, thousands of quasars have been identified, giving off energy at X-ray and infrared wavelengths as well as light and radio waves. They are all remote, and some are the most distant objects we know in the Universe. The quasar PKS 2000-330, for example, is at a distance of 13 billion light-years. This means that we are seeing it as it was at a time soon after the Universe was born.

ABOVE
**Gamma-ray burster**
The HST spots the optical counterpart (arrowed) of an intense gamma-ray burst (GRB 970228) in the outer reaches of a remote galaxy. Astronomers reckon that the collision between two neutron stars might have triggered the violent release of energy.

LEFT
**Quasar host galaxies**
The HST has targeted many quasars, revealing details of the host galaxies. Some galaxies seem undisturbed by quasar activity. In others, the quasars seem to be fuelled by the debris from collisions between different galaxies.

**RIGHT**
**Spiral jet**
The HST's Advanced Camera for Surveys (ACS) has spotted a giant radio jet coming from a spiral galaxy (0313-192), seen at far right. Right, the ACS image is shown superimposed on a radio image from the Very Large Array radio telescope. It is the first time that a radio jet has been spotted coming from a spiral galaxy.

**BELOW**
**Blowing bubbles**
A black hole lurks at the center of NGC 4438, a member of the Virgo cluster of galaxies. Here, the HST has spied glowing bubbles of hot gas rising from the accretion disk surrounding the black hole.

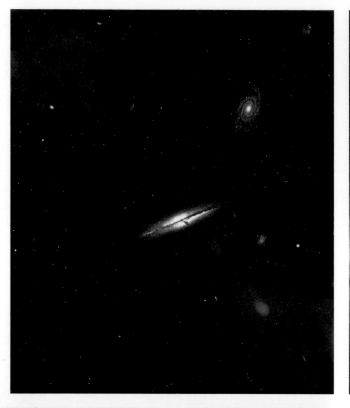

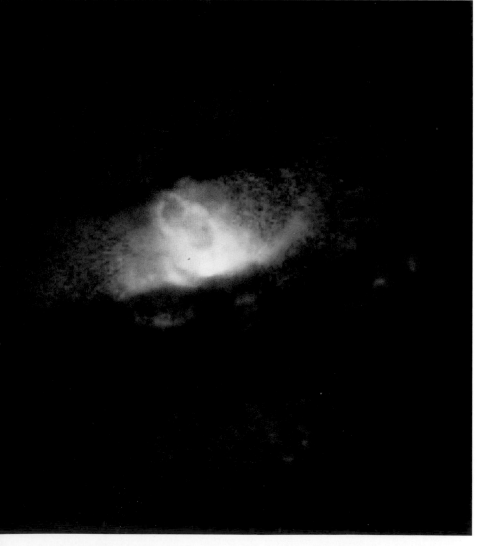

## VARYING IN BRIGHTNESS

Quasars appear to be a very compact source of energy, much smaller than ordinary galaxies. But just how big are they? We can find out because of how their brightness varies. The brightness of a quasar can fluctuate on a timescale of a day or less.

But it is a fact that the shortest time in which an object can vary in brightness is the time it takes for light to travel across it. This means that a quasar that varies in brightness in a day must be only about a light-day across. This makes it comparable in size to our solar system.

## BLAZARS

A blazar is another kind of active galaxy that varies widely at optical wavelengths. The name comes from a contraction of the terms BL Lacertae and quasar. BL Lacertae is an object first classified as a variable star but later found to be a strong radio source. It is notable because it does not show any lines in its spectrum.

## THE ENERGY MACHINE

Clearly, for active galaxies to be able to pump out such prodigious energy, they must have a powerful "engine." Astronomers reckon that the only engine powerful enough is a supermassive black hole.

Black holes are so-called because they have such powerful gravity that nothing can escape from them—not even light. Relatively small black holes are created when big stars die (see page 62), but supermassive black holes are formed by the collapse of matter in the center of youthful galaxies.

Black holes produce energy when matter is sucked into them. But matter does not simply travel straight down. Because of a galaxy's rotation, matter is strung out into an accretion disk. Surrounding the disk is a donut-shaped torus (ring) of cooler gas and dust.

The accretion disk rotates rapidly and gets so hot that it emits X-rays and other forms of electromagnetic radiation. The radiation can't escape through the disk, and is instead beamed along the rotating axis. Subatomic particles produced in the whirling disk also pour out along the axis, forming high-speed jets. The jets give off radio waves as they encounter particles in the surrounding space.

## POINT OF VIEW

Astronomers believe that the various kinds of active galaxies are different views of the same black-hole structure. When our line of sight is along the plane of the disk, the cool torus obscures the bright center. But we can detect the radio-emitting regions on each side, created by the jets smashing into the intergalactic medium, and so we see a radio galaxy. When the disk is angled more toward us, we see a quasar or a Seyfert galaxy. And if the disk appears face-on, with a jet pointing directly at us, we see a blazar.

ABOVE

**Energy transfer**
Fine, red thread-like structures are visible in this Hubble image of the giant elliptical galaxy NCG 1275. They are composed of relatively cool gas being suspended by a magnetic field. They begin to form when the cool gas is being transported from the supermassive black hole in the galaxy's center out to the surrounding gas.

# CENTAURUS A

Galaxy NGC 5128 is one of the brightest galaxies in the heavens and only just beyond naked-eye visibility. It is not included in Charles Messier's catalogue of clusters and nebulae, and therefore, does not have an M number.

This was not an oversight on Messier's part, but a reflection on the galaxy's location deep in southern skies and beyond Messier's reach. It lies in the constellation Centaurus (the Centaur) just north of Crux (the Southern Cross).

English astronomer John Herschel (son of William, who discovered Uranus) was one of the first European astronomers to study the southern skies, from South Africa, in the 1830s. He declared the galaxy to be "a most wonderful object, cut asunder by a broad, obscure band."

And NGC 5128 is a strange, beautiful object when viewed through a telescope. It is spherical in shape and is bisected by dark dust lanes. It lies about 15 million light-years away. In 1949, radio astronomers in Australia found a powerful radio source in Centaurus, which they called Centaurus A. They soon identified it with NGC 5128.

Astronomers speculated on the nature of this peculiar bisected galaxy: Could it be a spiral galaxy in the making, and the dark lanes part of a dusty disk that will eventually form spiral arms? Or could its appearance result from a collision between two galaxies—an elliptical and a spiral? Some astronomers favored the latter, reckoning that the radio emission was generated by the collision.

We now know that NGC 5128, usually called Centaurus A, is an active galaxy. It is the nearest active galaxy to us and the third most powerful radio source in the heavens (after Cassiopeia A and Cygnus A). It has been intensively studied by the most powerful radio telescopes and at other wavelengths by the HST, which peers deep into its mysterious dark dust lanes.

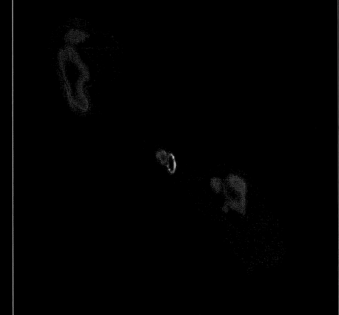

**ABOVE**
**On the radio**
This radio image of Centaurus A shows the typical wide-spaced lobes of radio galaxies. They extend about 20,000 light-years on either side of the galaxy's visible center.

**BELOW**
**Peering into the dust**
In the HST image at left, clusters of hot blue stars line the edges of the dust lane, where star formation is most intense. An infrared view (right) shows a swirling mass of hot gas, caught up in the gravitational whirlpool of a large black hole.

**RIGHT**
**The bright center**
In visible light, Centaurus A is one of the most intriguing and unmistakable objects in southern skies. The picture shows the bright central region of the galaxy, measuring about 30,000 light-years across, only about a fifth of its true extent.

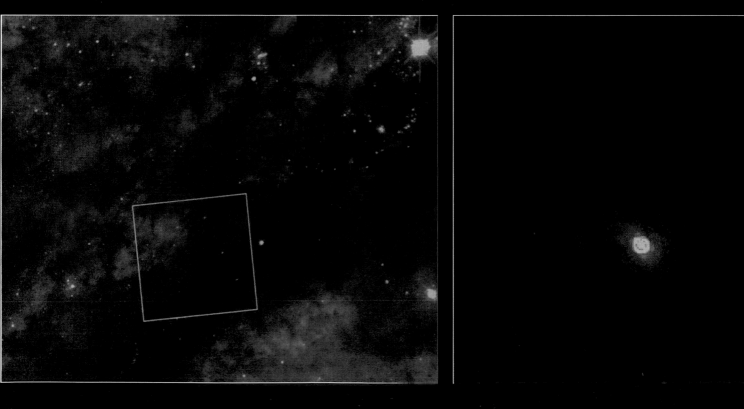

# GALAXIES IN COLLISION

When most people think about the Universe they visualize a huge expanse of space populated by widely separated galaxies. And when told that the Universe is expanding, they imagine the galaxies getting gradually farther and farther apart. The idea of galaxies colliding doesn't spring to mind.

In reality it is not individual galaxies but clusters and superclusters of galaxies that are moving apart. Individual galaxies stay within their cluster, where each one follows its own elliptical orbit around the center of mass of the cluster. As they orbit, the galaxies can make close encounters and so near misses and galactic collisions occur every now and then.

In fact, over the 13.7 billion year lifetime of our Universe, many galaxies have been in collisions. Looking deep into space, we find many examples. But even when we look at galaxies that are close to us, in the present time, we can see collisions taking place.

## GALACTIC ENCOUNTER

Galaxies are not solid objects and so unlike dodgem cars they don't just bump and bounce apart. They are made of millions or billions of stars, as well as large amounts of gas and dust. And these stars are vast distances from each other.

The typical distance between stars of our Milky Way Galaxy is about 50 million times larger than the typical diameter of an individual star. If our galaxy were to hit another galaxy, the two bodies would essentially go through each other. Only a handful of individual stars would actually smash into each other. The vast majority of them would pass each other by.

The two galaxies would eventually part but they would not be unaffected by the encounter. Gas and dust clouds of the two galaxies would intermingle, and energy would be lost. And the overall shapes of the two would change; the spiral structure of the Milky Way would be distorted. Due to energy loss, the relative velocity between the two galaxies would have decreased. Some stars might even have been flung out of each galaxy to form long stellar streamers.

Galactic collisions such as this occur very slowly; typically it takes several hundred million years for one galaxy to pass through another. And that encounter could just be the start of things. Later, the two galaxies could collide again, and then again, until eventually they merged into a single more massive and larger galaxy.

## COLLISION SPEED

Galaxies where stars are thinly spread are less affected by collisions than galaxies with stars clumped together. Velocity also plays a part. But interestingly, unlike motor car collisions, the slower the intercept speed the more "damage" is done. If galaxies collide quickly they spend relatively little time "inside" each other and so gravitational interaction has less time to work. Slow encounters are much more effective in producing galactic mergers and alterations.

A few hundred million years ago, a high speed (many hundreds of miles/kilometers per second) collision took place

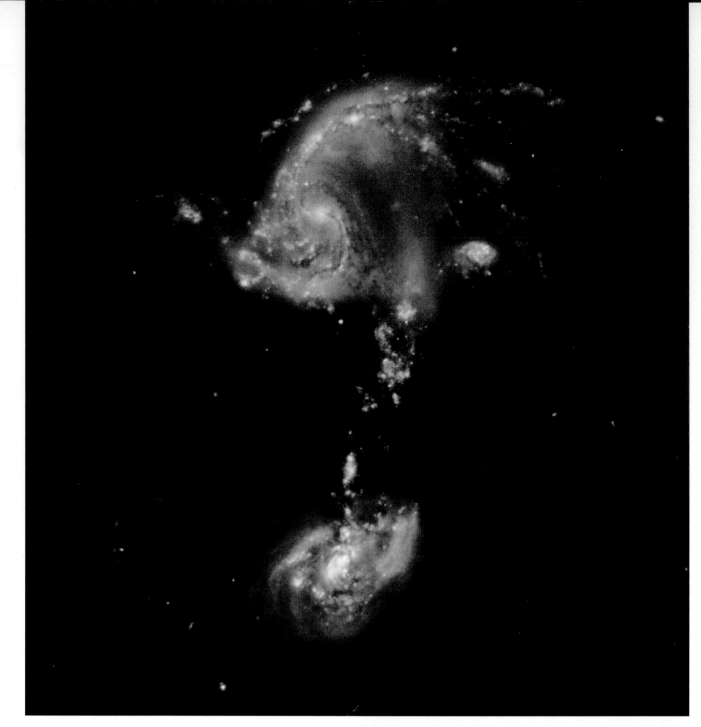

LEFT
**Blue streamer**
This 2009 Hubble image is of an interacting group of galaxies called Arp 194. At least two galaxy cores are merging in the top part of the view. A third, relatively normal spiral is to the right. Extending below this group is a blue stream made of millions of massive, young stars created as a result of the merger. The stream lies in front of a larger spiral galaxy.

RIGHT
**Head-on collision**
The Cartwheel Galaxy is the result of a head-on collision between two galaxies. In the distant past a small galaxy moved through the core of the previously normal spiral galaxy. This sent a ripple of energy into space triggering a burst of star formation. The result is the cartwheel shape which consists of a central core and an outer ring made of several billion new stars.

between galaxies NGC 4038 and NGC 4039. This encounter produced the Antennae galaxies that we see today just 45 million light-years away. A pair of long, curving streamers of stars, gas, and dust resembling the antennae of an insect, curl away in opposite directions from the merging system.

## SIZE MATTERS

Often small galaxies hit large ones. As the small galaxy approaches and passes through the large one it becomes distorted into an irregular shape. This happened to both the Small and Large Magellanic Clouds (see page 80) Both of these irregular galaxies have passed through our own Milky Way and are today satellites orbiting our Galaxy.

Big galaxies tend to accrete satellites as time progresses. Depending on the orbits of these satellites around the larger galaxy, this can lead to the production of galactic cores that rotate in the opposite direction to their outer disks. And on other occasions this leads to galaxies with double cores and with disks that are featureless and bland.

## TRIGGERING STAR FORMATION

Another effect of collisions is the compression of some of the giant molecular clouds in a galaxy's disk. This leads to a burst of star formation. New massive stars quickly reach the supernova stage of their lives. Their supernova explosions can lead to supersonic winds that expel even more gas from the gravitational grip of an individual galaxy.

These star-burst galaxies emit bluer light than isolated galaxies of a similar type. They produce about 1,000 stars each year, In comparison with the two to three In the MIlky

RIGHT
**Cosmic collisions**
These five images were among 59 views of colliding galaxies that were released to the public in celebration of Hubble's 18th launch anniversary in 2008. The images capture merging galaxies in various stages of collision and show the variety of structures they form; in the process providing valuable information on the evolution of galaxies.

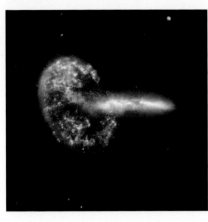

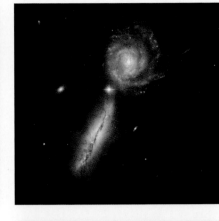

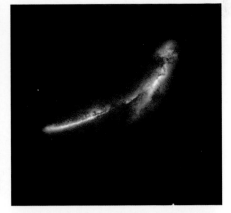

Way Galaxy. This happened spectacularly in the aptly named Cartwheel Galaxy, with vigorous star formation occurring at the "rim" of the wheel. This galaxy was once rather like the Milky Way but it apparently underwent a head-on collision with a smaller companion galaxy about 200 million years ago. Another good example of starburst activity is M82 (see page 67) just 12 million light-years away. This galaxy is being disturbed by the larger nearby spiral galaxy M81.

Galactic collisions can also lead to warped galactic disks. Radio surveys of the disks of galaxies, and especially the distribution of neutral hydrogen atoms in those disks, indicate that about fifty percent of all galactic disks are not flat but warped. Once the warping has been produced it can last for 3 to 5 billion years.

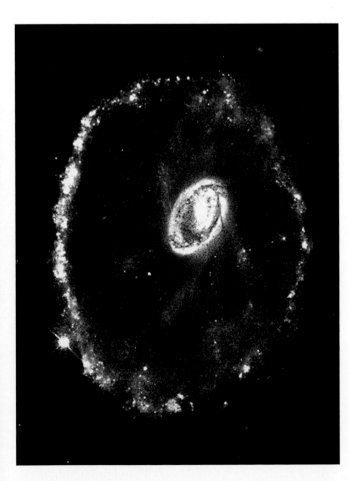

## MERGERS AND TAKEOVERS

When a small and a large galaxy interact the larger one generally eventually swallows up the smaller one, and thus grows in mass and size. Over time this will lead to the Universe containing fewer but more massive, large galaxies.

These mergers also tend to convert spiral galaxies into larger elliptical or irregular galaxies. Hubble's investigation of the dense Coma Cluster of galaxies shows that the ratio between the number of spiral galaxies and the number of elliptical and irregular galaxies varies depending on the distance from the center of the cluster. Elliptical and irregular galaxies together dominate the central region, a region where collisions are more prevalent.

One easy proof that galaxies collide is the fact that we can see five pairs of galaxies actually colliding with each other in the nearest 300 million light-years to us. The Hubble telescope has also imaged nearly a hundred other colliding galaxies and this has produced a fascinating insight into the time sequence of the changes that occur.

What is even more important is Hubble's ability to detect faint and very faint galaxies in its deep-field and ultra-deep-field images (see page 202). Here we look ever further back in time towards the dawn of the Universe. The suggestions that galaxies were smaller then, because fewer collisions had taken place can be experimentally tested. We can also investigate if the ratio between the number of spiral galaxies and the number of giant elliptical and irregular galaxies has varied over time.

FOLLOWING PAGES
**Galaxy mask**
A giant galactoc mask has been produced as a result of a close encounter between two galaxies; NGC 2207 (left) and smaller IC 2163. Strong tidal forces from NGC 2207 have distorted IC 2163's shape, flinging out long streamers of stars and gas. The two met about 40 million years ago and will eventually merge into one huge galaxy.

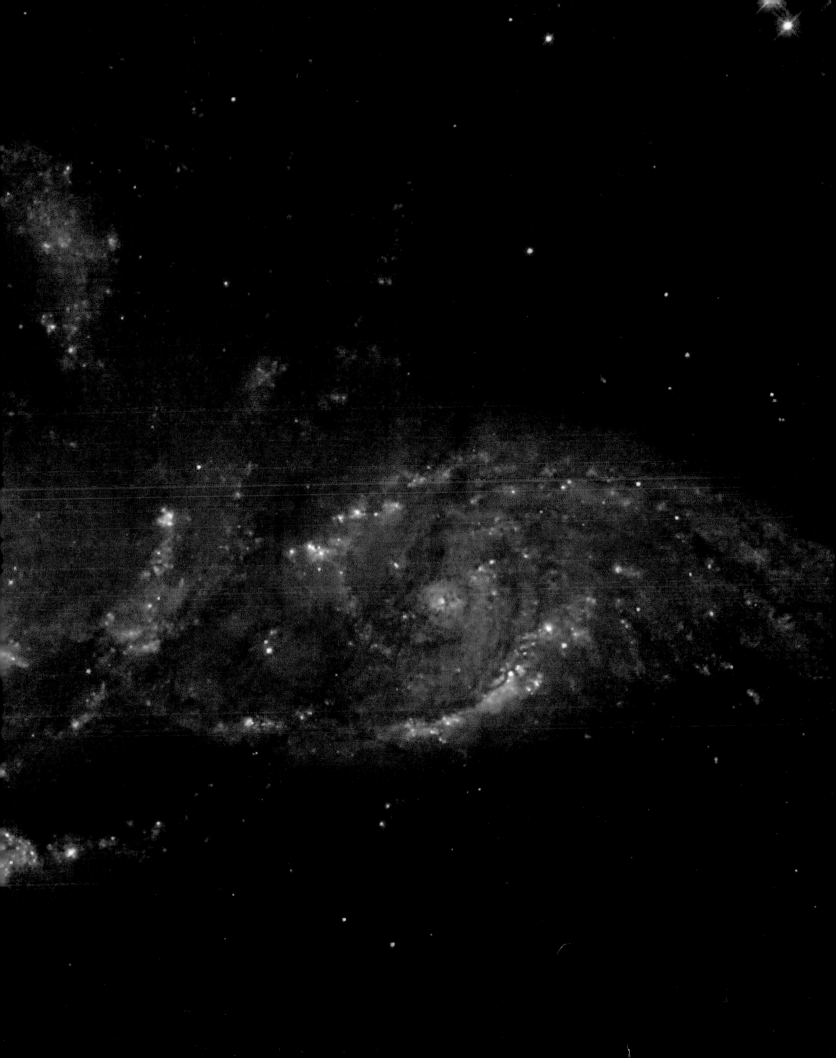

# IN THE WHIRLPOOL

William Parsons, the third Earl of Rosse, was an Irish landowner, whose home was at Birr Castle, Parsonstown, not far from Athlone in central Ireland. Upon graduating from Oxford in 1822, he first dabbled in politics, but abandoned that for astronomy in 1834.

Not content to buy a telescope, Lord Rosse decided to build his own. By 1838, he had built a reflector with a 36-inch (91-cm) mirror. It was an excellent instrument which encouraged him to attempt one twice as large. No one had ever tried to build one this size before. In 1845, he completed the 72-inch (1.8-m) reflector. It was a monster. The mirror was housed in an iron tube nearly 60 feet (18 m) long and 8 feet (2.4 m) across. It was mounted between two massive stone walls and was aptly named the Leviathan of Parsonstown.

With his new telescope, Lord Rosse decided to start observing the star clusters and nebulae that Charles Messier had listed in his catalogue. When he turned the Leviathan on the nebula with the Messier number 51, he was astonished to find that it was shaped like a pinwheel. It had "spiral convolutions," he said. He was the first to discover the spiral nature of the nebulae that would in the 1920s be recognized as external galaxies.

M51 lies about 20 million light-years away in the constellation Canes Venatici (the Hunting Dogs). It is easy to find with powerful binoculars or a small telescope, because it lies just south of Alkaid, the first star in the handle of the Big Dipper, which is in the neighboring constellation of Ursa Major (the Great Bear). Its nickname, the Whirlpool Galaxy, is apt because we see it face-on, with its spiral arms beautifully revealed.

M51 is actually not one galaxy, but two. The main spiral is NGC 5194, and one of its arms is tenuously linked with a smaller galaxy, NGC 5195. The present structure has resulted from a glancing collision between the two galaxies that probably took place around 300 million years ago. The smaller galaxy was by then an ordinary spiral, or more likely a barred spiral. As it slammed into the outer regions of the large spiral, NGC 5195 was torn apart. Most of its stars were strung out to form a bridge with the other galaxy.

For its size, M51 is particularly bright, with an abundance of young, hot blue stars. Though only about two-thirds as big across as the Milky Way, it is three to four times brighter. Powerful bursts of radio waves come from the centers of both galaxies. The exceptional activity taking place in them almost certainly stems from the collision, as usually happens when galaxies interact.

RIGHT
**Heart of the Whirlpool**
On the HST's 15th anniversary in April 2005 it returned to the celebrated Whirlpool Galaxy. The two curving arms are shown in striking detail. Clusters of young stars are highlighted in red.

BELOW LEFT
**Bridging the gap**
This wider-angle ground telescope image shows well the Whirlpool's widespread spiral arms and the bridge of gas between it and its companion.

BELOW RIGHT
**Core of the spiral**
This Hubble view of the spiral galaxy's core reveals a dark "X" shape. It marks the exact position of a central black hole. The X is formed by dust; each bar of the X is thought to be an edge-on dust ring surrounding the hole.

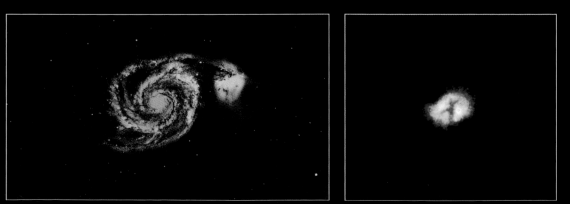

# 4 | The Expansive Universe

### The Universe might continue to expand forever

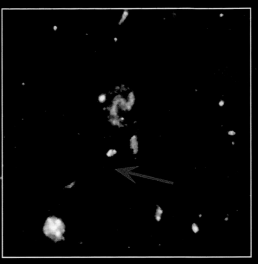

**ABOVE**
**Cosmic magnifier**
The gravity of a cluster of galaxies has worked like a lens and bent the light from a more distant quasar. Five separate images of the quasar are produced surrounding the cluster's center. The stretched arcs are more distant galaxies lying beyond the cluster.

**INSET LEFT**
**Heart of a globular**
The HST looks into the heart of the globular cluster M4, which is 5,600 light-years away in the constellation of Scorpius. It has identified individual stars, such as white dwarves and pulsars, among the 100,000 plus old stars within the cluster.

**INSET RIGHT**
**Standard candle**
With its phenomenally sharp vision, the HST helps pinpoint "standard candles" in remote galaxies. These objects help astronomers make more accurate estimates of the scale of the Universe.

# AFTER THE BIG BANG

Everything that exists—our bodies, the air we breathe, the Sun, the stars, the galaxies, and space itself—makes up the Universe. From year to year we see our local neighborhood, the solar system, change markedly. Changes in the stellar heavens and in the Universe at large are much more subtle, but they do occur.

The Universe is changing and evolving. But where did it come from? How has it evolved? Exactly how is it made up? And what will happen to it in the future? The astronomers who seek to answer such fundamental questions are known as cosmologists. Cosmology is the study of the origins and evolution of the Universe.

Irishman James Ussher, who became archbishop of Armagh in 1625, was no cosmologist, but using biblical references he calculated that the creation of the Earth and the heavens took place on an October morning in the year 4004 BCE. He was far off—by a factor of nearly a million. Studies of radioactive rocks suggest that the Earth, along with the rest of the solar system, was born about 4,600 million years ago.

The Universe is about three times older. Astronomers believe that it came into being 13 to 14 billion years ago. How do they know this? Well, they find that the galaxies are all rushing away from us and from one another. The Universe seems to be flying apart, expanding, as though from a massive explosion long ago.

And astronomers reckon that a kind of explosion really did take place—they call it the Big Bang. By measuring the rate at which the Universe is expanding now, and calculating backwards, they have determined when the explosion happened—when the Universe was created. It was about 13.7 billion years ago.

What happened before the Big Bang? Today's cosmologists think this question is impossible to answer because when the Big Bang happened, time itself began.

Cosmologists think they know what happened after the Big Bang—except for the first $10^{-43}$ seconds. During this period, called Planck time, space and time itself, the fundamental forces of nature, and the laws of physics were still forming. After this time, cosmologists can apply knowledge of fundamental forces and laws to describe what happened next.

The Universe remained tiny until it was $10^{-35}$ seconds old. It was incredibly hot (around 1027 degrees Kelvin) and full of energy. Gravity had already become a distinct force, and the first fundamental particles, including electrons, began to form. The Universe then expanded suddenly, in an event called inflation. As the Universe ballooned, its temperature dropped rapidly.

When the Universe was one-millionth ($10^{-6}$) of a second old, protons and neutrons started to form. Protons are the nuclei (centers) of hydrogen atoms. By about three minutes, the temperature of the Universe was only about one billion (109) degrees Kelvin. Protons and neutrons began to stick together, forming the nuclei of deuterium (heavy hydrogen) and helium. At this time the proportions of hydrogen and helium in the Universe were fixed.

For about the next 300,000 years, things quietened down somewhat. The Universe continued expanding and cooling, but it stayed essentially the same—full of radiation and a mass of atomic nuclei in a sea of electrons. It was a "foggy" Universe. Because of all the particles milling around, radiation couldn't travel far without being scattered, like light in a fog.

About 300,000 years after the Big Bang, temperatures fell to about 5,000 degrees Fahrenheit (3,000°C). Electrons could now combine with nuclei to form the first atoms of hydrogen and helium. The "fog" cleared as particle numbers were drastically reduced, allowing radiation to pass relatively unhindered. The Universe became transparent.

Cosmologists estimate that some time within the first one or two billion years, hydrogen and helium began to form clouds. These would later collapse and create the first stars and galaxies. The HST has made a systematic search for the first bright galaxies to form in the early Universe. Hundreds of bright galaxies have been found at around 900 million years after the Big Bang, but very few before then.

**RIGHT (MAIN IMAGE)**
**Galaxy evolution**
Thousands of galaxies at various stages of evolution are captured in this deep-sky view. Mature spirals and ellipticals are in the foreground. Smaller, fainter, irregular-shaped galaxies are more distant and so further back in time. These smaller ones are the building blocks of the larger galaxies of today.

**RIGHT INSETS**
**Young galaxies**
Astronomers studying HST data have found more than 500 galaxies that existed within a billion years of the Big Bang. They appear red because of their tremendous distance from Earth. During the journey to Earth, the blue light from the galaxies' young stars shifted to red light due to the expansion of space.

**RIGHT**
**First stars**
The Universe's first stars have not been detected yet but we know they formed from clouds of hydrogen and helium. Once generating energy, the hot young stars would illuminate the surrounding clouds, in the way recently formed stars do here.

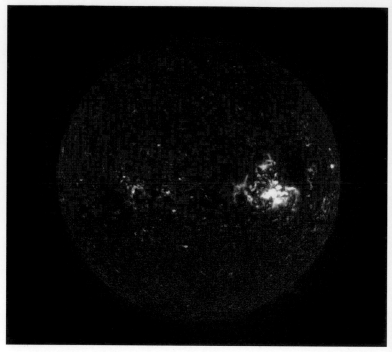

# THE SCALE OF THE UNIVERSE

How big is this Universe of ours? Unimaginably vast. But this is a relatively recent realization. Early astronomers knew that the Universe was bigger than Earth (which they thought was its center), but they did not believe it was much bigger. Later, astronomers coming to terms with Copernicus's concept of a solar system, equated this solar system with the Universe.

Starting in William Herschel's time, in the eighteenth century, people considered the Milky Way Galaxy to be the whole Universe. In 1918, U.S. astronomer Harlow Shapley calculated the Milky Way's size to be 300,000 light-years; it turns out this estimate was three times too big. But the true scale of the Universe did not became apparent until five years later, when Edwin Hubble proved that spiral "nebulae" were actually separate and remote galaxies, millions of light-years away.

Since then, astronomers have progressively pushed back the boundaries of the Universe, using ever-more-powerful telescopes and, more recently, space telescopes like the HST.

## FROM ATOMS TO SUPERCLUSTERS

The tiniest objects that exist in the Universe are particles within atoms—the basic constituents of ordinary matter. Atoms measure about $10^{-10}$ meters across, or 1/10,000,000,000th of a meter (atomic sizes are expressed as fractions of a meter, never in fractions of an inch). This means that it would take 10 billion atoms, side by side, to measure a meter. A hundred thousand times smaller still, at around $10^{-15}$ meters across, are the electrons that circle the atomic nucleus. They are among the smallest of all atomic particles, slightly bigger than particles known as quarks ($10^{-16}$ meters).

We need to scale things up a billion times or so to bring us into the realm of our everyday life, when measurements are made in inches, feet, yards, centimeters, meters, and so on. A few million times bigger is the Earth itself, at about $10^7$ meters. And a billion times bigger still takes us way beyond the planets and a quarter of the way to the nearest star beyond the Sun.

When the numbers get this large, measurements become difficult to understand, so we switch to expressing distances in light-years: one light-year is about $10^{12}$ miles ($10^{13}$ km). A quarter of the distance to the nearest star is thus about one light-year.

Multiply this distance by 100,000, and we get the size of the next-biggest entity, the Milky Way Galaxy. Multiply it 50 times more, and we're up to the scale of the cluster of galaxies to which the Milky Way belongs. Twenty times more, and we're looking at a supercluster. Two-hundred times more, and we are more than 10 billion light-years away and nearing the edge of the observable Universe.

This distance works out at $10^{26}$ meters. The scale of the Universe—from the smallest object to the largest, from a quark to the observable edge of space—goes from $10^{-16}$ to $10^{26}$ meters. In other words, the Universe is more than a million, million, million, million, million, million, million ($10^{42}$) times bigger than the smallest particle.

ABOVE LEFT
**Planet Earth**
Our verdant home in space, where conditions are ideal for life to thrive in abundance.

ABOVE RIGHT
**The Sun**
Our local star, which pours heat, light, and other radiation into space—we see it here through the eyes of the SOHO spacecraft.

RIGHT INSET TOP
**Stars**
The far-distant suns that shine in the night sky, as here in the constellation Cygnus, one of the few that lives up to its name (the Swan).

RIGHT INSET BOTTOM
**A galaxy**
The spiral galaxy M81, like other galaxies, is a huge star island in space populated by billions of stars.

RIGHT (MAIN IMAGE)
**The deep Universe**
Galaxies galore appear in deep space images. This typical patch of sky imaged by the HST includes galaxies, far and near, big and small.

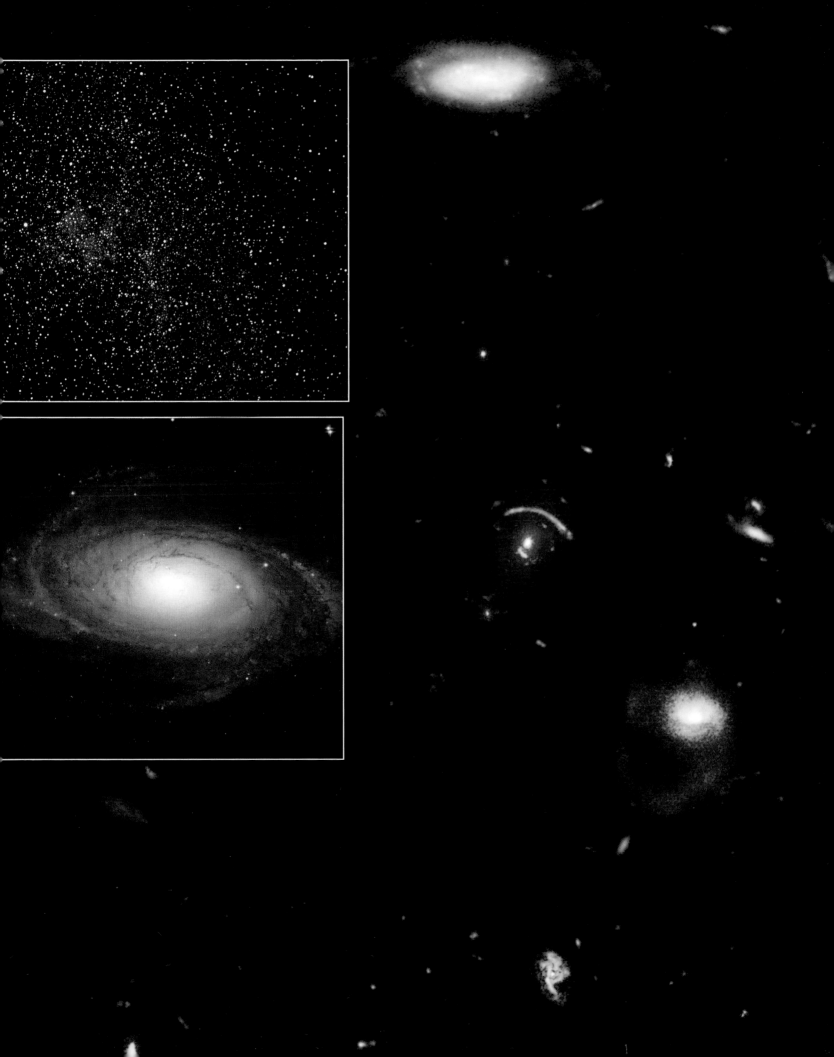

# THE HUBBLE ULTRA DEEP FIELD

One of the most celebrated and astronomically important images that the HST has produced is the Hubble Ultra Deep Field (HUDF). It is the deepest ever view into the visible Universe and reveals galaxies that are too faint to be seen using Earth-based telescopes. Looking at the image we are looking back in time to when the first galaxies took shape after the Big Bang event that created the Universe.

The HUDF image is of a tiny patch of sky in the constellation of Fornax. A region so small that fifty patches this size would be needed to cover the full Moon. The view, which appears virtually empty in Earth-based images, has been compared to looking through an 8-foot (2.4-meter) drinking straw.

An estimated 10,000 galaxies of various sizes, shapes, and colors have been captured in the image. There are spirals and ellipticals as well as many oddly shaped galaxies; some of which are obviously interacting. The view reveals a Universe more chaotic than the ordered, closer Universe of today.

The image is actually two separate Hubble images. One was taken with the Hubble Advanced Camera for Surveys (ACS); the phone-booth-sized instrument that was installed on the 2002 service mission. It took 800 exposures over the course of 400 Hubble orbits around Earth. Work began on September 24th 2003 and continued until January 16th 2004. The total exposure time was about 1 million seconds (11.3 days). To image the entire sky in this way would take almost one million years of continuous observing. The ACS uncovered galaxies that existed 800 million years after the Big Bang.

The ACS view was combined with the second image taken by the Near Infrared Camera and Multi-Object Spectrometer (NICMOS) which could see objects at greater distance and so uncovered even younger galaxies. The photons of light from the galaxies they viewed began their journey across the Universe before Earth existed. They arrived at the rate of about one photon per minute compared to the millions per minute from closer galaxies.

The HUDF patch of sky was observed once again by Hubble in 2009–10. This time by the Wide Field Camera 3 (WFC3), installed during the 2009 service mission (see page 204). The image exposure lasted a total of 87 hours and was taken between August 26th 2009 and September 14th 2010. The camera's near-infrared eyes looked deeper into space than

(see page 204)

**RIGHT**
**Deepest view**
The Hubble Ultra Deep Field is the deepest portrait of the visible Universe so far. In this 2009–10 view we are taken back to the early Universe when the first galaxies were emerging. The most distant objects and so those furthest back in time appear red because their light is stretched to longer, redder wavelengths.

**BELOW**
**Looking back**
Hubble has looked further and further back in time towards the Big Bang. New cameras installed after servicing missions meant that in 2004, and then in 2009–10 it could see younger and younger galaxies. The Webb telescope is expected to look even further back; possibly to the first stars.

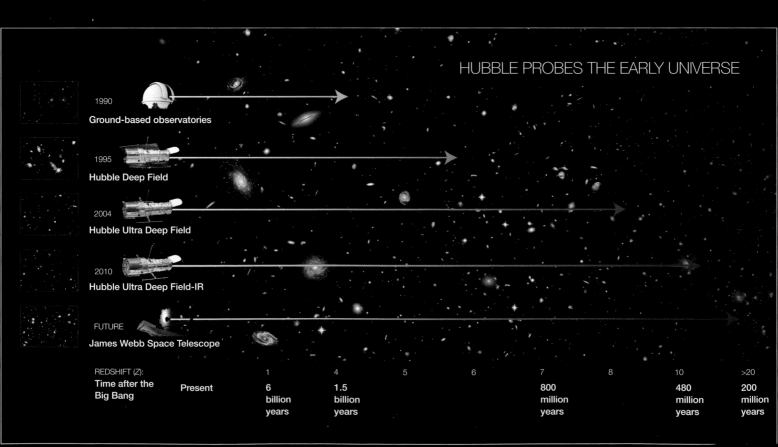

## HUBBLE PROBES THE EARLY UNIVERSE

1990
**Ground-based observatories**

1995
**Hubble Deep Field**

2004
**Hubble Ultra Deep Field**

2010
**Hubble Ultra Deep Field-IR**

FUTURE
**James Webb Space Telescope**

| REDSHIFT (Z): | | 1 | 4 | 5 | 6 | 7 | 8 | 10 | >20 |
|---|---|---|---|---|---|---|---|---|---|
| **Time after the Big Bang** | Present | 6 billion years | 1.5 billion years | | | 800 million years | | 480 million years | 200 million years |

even NICMOS. In January 2011 the youngest galaxy in the view was identified. It is a small compact galaxy of blue stars called UDFj-39546284. It is estimated at a mere 480 million years after the Big Bang, and it had started to form some 100–200 million years earlier.

Hubble had peered into deep space twice before it took the HUDF. The two previous ground-breaking views were the Hubble Deep Field of 1995, which looked at a small region within the constellation of Ursa Major, and the Hubble Deep Field South of 1998 that studied a small region in Tucana.

The HUDF will remain the deepest view into the Universe until the James Webb Space Telescope (see page 208) is launched. Its infrared vision will investigate ultra-deep space with greater sensitivity; hunting out primeval galaxies at greater distance, and taking us even further back in time.

# COSMOLOGICAL DISTANCING

Proxima Centauri, the nearest star to the Sun, lies about 4.3 light-years away from Earth. Sirius, the brightest star in the heavens, lies 8.8 light-years away. The Andromeda Galaxy, the farthest object we can see with the naked eye, lies 2.9 million light-years away. The most distant galaxies we can see are about 13 billion light-years away.

How do we know that these figures are accurate? How do we measure the distance to the stars and galaxies? To measure the distance to a few hundred of the nearest stars, astronomers use a variation of the technique surveyors use to measure distances on land.

In this process, called triangulation, a surveyor first measures out a baseline—a straight line along the ground—and then measures the angles from each end of the baseline to a distant point, for example, a church. Knowing the length of one side (the baseline) and two angles, the surveyor can use simple trigonometry to calculate the lengths of the other two sides of the triangle formed by the ends of the baseline and the church. This pinpoints the location of the church.

## USING PARALLAX

For the astronomical version of triangulation, astronomers choose a much longer baseline: the diameter of the Earth's orbit around the Sun. They view a star from one side of Earth's orbit in January, for example, and then from the other side six months later. A nearby star seems to change position slightly against the background of distant stars during this time. This

RIGHT
**Measuring distances**
The spiral galaxy NGC 3021 is home to Cepheid stars and to a supernova that exploded in 1995. Locations of the Cepheids are highlighted by green circles within the four inset boxes. Observations of the Cepheids and the supernova were used to measure intergalactic distances and also to refine the Hubble Constant.

BELOW
**M100**
The remote spiral galaxy M100, one of thousands in a mammoth cluster of galaxies in Virgo. The HST can pinpoint individual stars in the spiral arms, and in particular Cepheid variables, which can be used as standard candles.

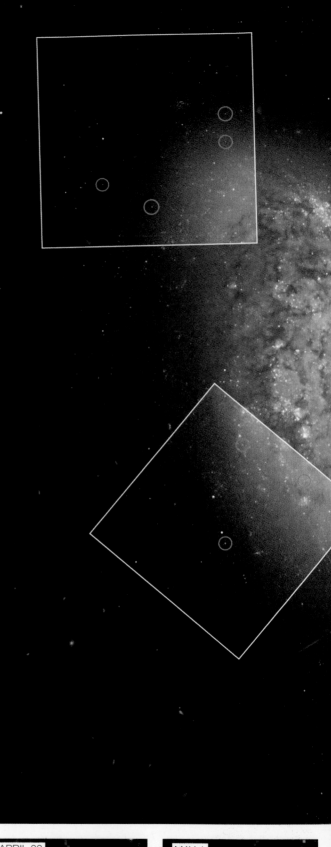

APRIL 23

MAY 4

Site of
SN 1995al

effect is called parallax. You can easily demonstrate parallax yourself. Hold up a finger in front of you and look at it first with one eye and then the other. You'll notice that your finger seems to shift position against the distant background, and the nearer your finger is to your face, the more it seems to shift.

From the parallax shift shown by a star, astronomers calculate its distance in the same way a surveyor would for the church. However, parallax measurement has severe limitations. Parallax shifts are tiny, even for the nearest stars, and for stars farther than about 200 light-years, they are almost impossible to detect. In the early 1990s, however, the European astronomy satellite Hipparchos managed to measure parallaxes of around 120,000 stars out to a distance of about 500 light-years.

### STANDARD CANDLES

Parallax measurement is certainly not an option for measuring distances to other galaxies. To measure galactic distances, astronomers look for what are called standard candles. These are objects that are always of similar luminosity or behave in a predictable way.

The classic standard candles are variable stars called Cepheids. Early last century, U.S. astronomer Henrietta Leavitt discovered a fundamental law about Cepheids: that the period over which they vary in brightness is directly related to their luminosity.

This relationship, known as the period-luminosity law, makes Cepheids standard candles. When astronomers measure the period of a Cepheid, they know what its true brightness is. They then compare this true brightness with the star's apparent brightness in the sky. Because light fades in proportion to increasing distance, it is then easy to calculate how far away the Cepheid is.

It was Edwin Hubble who first used a Cepheid as a standard candle to measure the distance to the Andromeda Galaxy, proving that it lay beyond our own Galaxy (see page 184). The HST has extended his work, pinpointing Cepheid and supernova standard candles in some of the more distant galaxies.

LEFT
**Cepheid variations**
From HST observations of the light variations of a Cepheid variable in M100 over several months, astronomers find its period and hence its luminosity, or true brightness. Comparing this with its apparent brightness, they calculate that the Cepheid lies 51 million light-years away.

MAY 9

MAY 16

MAY 20

MAY 31

## OTHER STANDARD CANDLES

Globular clusters are other objects that can be used as standard candles. These globe-shaped masses of stars orbit the centers of galaxies, and the brightest ones have about the same luminosity. By comparing this with their apparent brightness in the sky, astronomers can estimate their distance.

Supernovae are useful as standard candles for even more remote galaxies. These exploding stars can become for a while as brilliant as an entire galaxy and can therefore be spotted at enormous distances. The most useful kind of supernova is the Type I, caused by the explosion of a white dwarf star (see page 57), because they all reach close to the same level of brightness.

## RED SHIFT

The most remote galaxies are too far away to detect any kind of standard candle. To determine how far away they are, astronomers revert to the invaluable astronomical technique of spectroscopy.

Just as astronomers can tell a lot about a star by examining the spectrum of its light (see page 32), they can also tell a lot about a galaxy from its spectrum. They see in the spectrum the characteristic dark absorption lines, which identify certain chemical elements. They can tell by the position of the lines in the spectrum whether the galaxy is moving toward or away from us.

BELOW
**Standard globulars** Globular clusters can make useful standard candles to measure distances to some galaxies. Here in the core of the giant elliptical galaxy NGC 1275 in Perseus, the HST resolves a number of globular clusters (blue). These are unusual because they consist of young stars, not old.

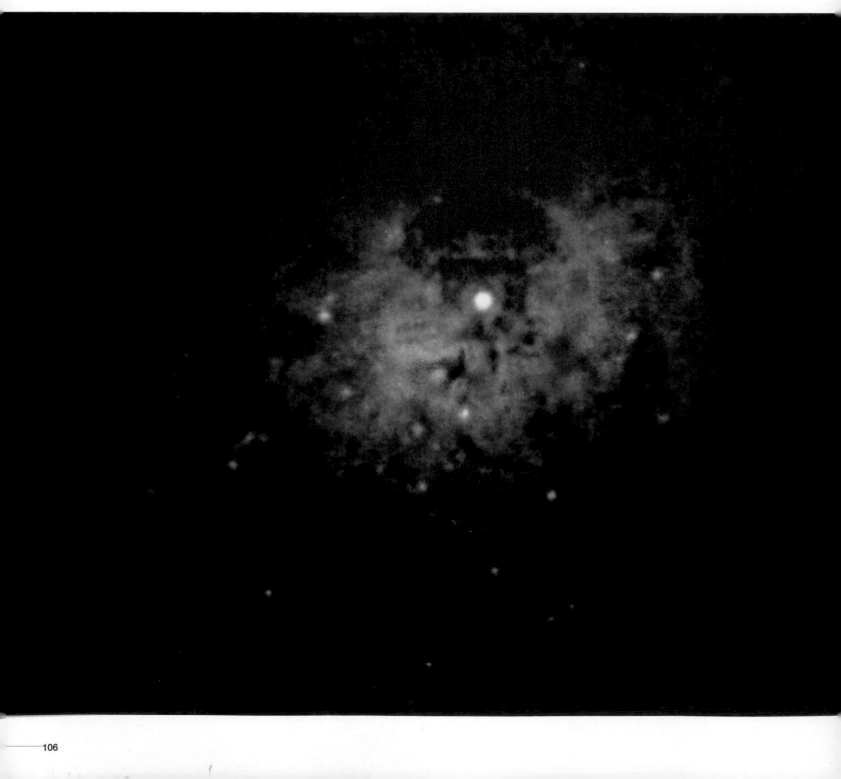

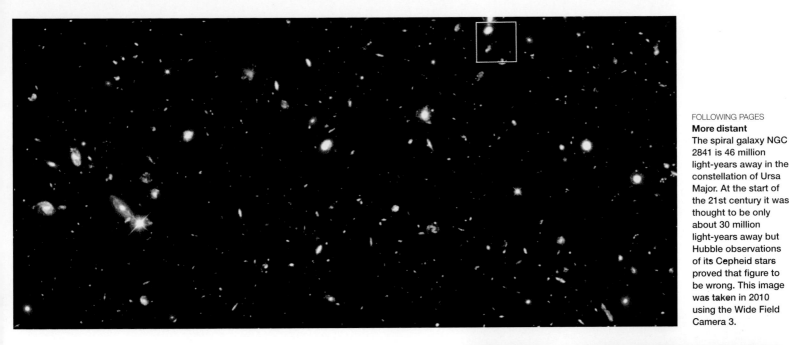

FOLLOWING PAGES
**More distant**
The spiral galaxy NGC 2841 is 46 million light-years away in the constellation of Ursa Major. At the start of the 21st century it was thought to be only about 30 million light-years away but Hubble observations of its Cepheid stars proved that figure to be wrong. This image was taken in 2010 using the Wide Field Camera 3.

When a galaxy is moving toward us, its light waves bunch up and seem to have a shorter wavelength, which makes them appear bluer. And the galaxy shows a spectrum with lines shifted toward the blue end. This is called blue shift. When a galaxy is moving away, its light waves stretch out and seem to have a longer wavelength, which makes them appear redder. And the lines in the spectrum shift toward the red end. This is called red shift. The extent of the blue or red shift tells us how fast the galaxy is moving.

This bunching up and stretching out of waves from a moving source is called the Doppler effect. We experience this with sound waves when an ambulance—siren blaring—races first toward us and then away. The noise of the siren has a higher pitch (shorter wavelength) when the ambulance is approaching, because the sound waves are bunching up. The noise has a lower pitch (longer wavelength) when the ambulance is going away, because the sound waves are being stretched out.

ABOVE AND RIGHT
**Ancient blast**
Comparing the light output of galaxies in the Hubble Deep Field over an interval of 2 years, HST scientists discovered in one of them a supernova (arrowed at right). The explosion had taken place 10 billion years ago. Supernovae prove useful in estimating distances to remote galaxies.

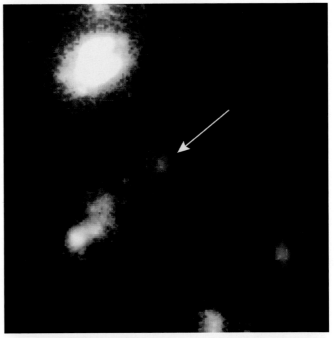

## RELATING RED SHIFT

Astronomers find that almost all the galaxies have a red shift, indicating that they are rushing away from us and from one another. This is how Hubble and other astronomers in the early twentieth century realized that the Universe must be expanding.

Hubble then made a discovery that is crucial to cosmological distance measurement. He found that the speed of recession of a galaxy (measured by its red shift) is directly related to its distance. This relationship is known as Hubble's Law. The rate at which galaxies speed up with increasing distance is called the Hubble Constant. Current calculations seem to show that a galaxy speeds up by around 45 miles (72 km) a second every megaparsec (about three million light-years).

By more accurately measuring the distance to remote galaxies, the HST has been helping refine the value of the Hubble Constant. The more accurate the value, the more accurately we can describe and comprehend the Universe.

RIGHT
**Spectral shifts**
The shift of the dark lines in the spectra of stars and galaxies tells us if they are traveling toward us (blue shift) or away (red shift). Spectral shifts can also detect rotations. Here, the HST's spectrograph has scanned the rotating disk around a black hole in the galaxy M84. The colors show the abrupt shift in wavelengths from one side of the disk to the other.

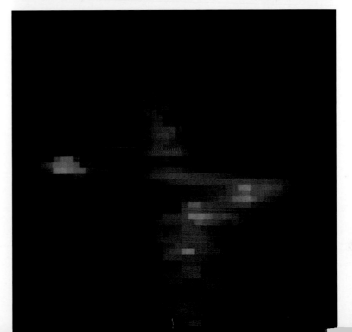

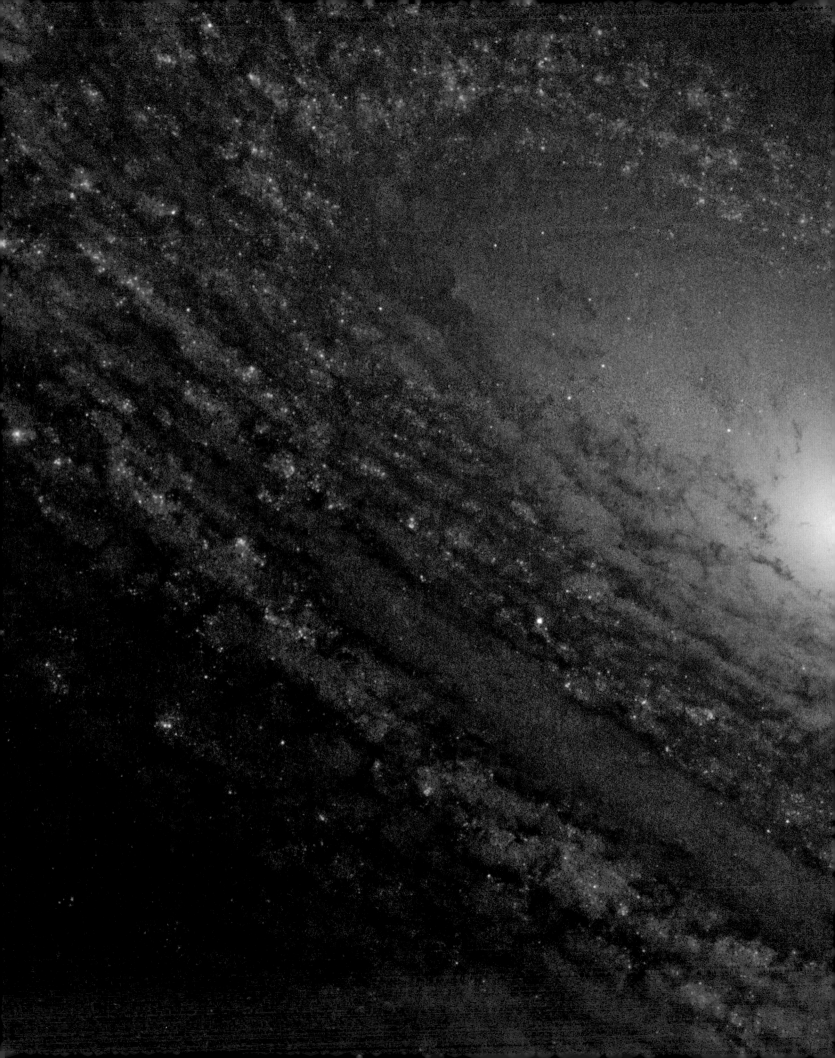

# GROUPING TOGETHER

Galaxies are not distributed evenly throughout space; they gather together into small groups or larger clusters. It is the mutual gravity of galaxies that helps keep them together and overcomes, on a small scale, their tendency to fly apart as the Universe expands.

In our corner of space, gravity draws together our Galaxy (the Milky Way) and about 40 other galaxies to form what is called the Local Group. The Milky Way is one of the two large spirals that dominates the Local Group. The other is the Andromeda Galaxy. There is just one other smaller spiral, M33. All the rest are small, dwarf elliptical, or irregular galaxies. Altogether, the Local Group occupies a region of space around five million light-years across.

## SATELLITE GALAXIES

Within the Local Group, smaller galaxies congregate around the two large spirals. We call them satellite galaxies. The two Magellanic Clouds we can see with the naked eye in southern skies are satellites of the Milky Way. Our Galaxy also has many other satellites, all of them dwarfs and all too far away to see with the naked eye.

The Sagittarius dwarf galaxy, about half as wide across as the Large Magellanic Cloud (LMC), lies even closer to us, at a distance of around 78,000 light-years. This is so close that the gravity of the Milky Way is ripping it apart. Obscured by the dust in the plane of the Milky Way, this galaxy was not discovered until 1994.

Since the Sagittarius dwarf galaxy is nearly spherical, it is classified as elliptical. Many of the Milky Way's other satellites are dwarf ellipticals. They include the diminutive Draco and Carina systems, both around 250,000 light-years away and both only about 500 light-years across—only 1/200th the size of our own Galaxy. They may contain as few as 300,000 stars.

## THE OTHER SPIRALS

The Andromeda Galaxy is a spiral, just like the Milky Way but considerably larger. This magnificent galaxy has figured prominently in the history of astronomy and is profiled overleaf.

The Andromeda Galaxy also has satellites orbiting round it. The third spiral in the Local Group, M33, seems isolated, although many astronomers consider it to be a satellite too.

M33 is called Triangulum Galaxy because it lies within the tiny northern constellation of that name, whose three brightest stars form the shape of a triangle. M33 looks magnificent from Earth because we see it face-on and have an excellent view of its spiral arms. It is right at the limit of naked-eye visibility and only those observers with exceptional eyesight might spot it under very clear conditions. However, it is easy to see with binoculars, visible in the same field of view as the constellation's lead star, Alpha. One striking feature of the galaxy is a giant nebula of hydrogen gas, NGC 604, which is a vast star-forming region on one of M33's spiral arms.

ABOVE
**Wide-open arms**
The wide-open arms of M33 are well seen in telescopes. The galaxy lies 2.7 million light-years away, a little closer than the Andromeda Galaxy. With a diameter of some 40 million light-years, it is less than half the size of our own Galaxy.

RIGHT
**Star-birth spectacular**
Rich populations of infant stars found in N90, one of the star-forming regions in the Small Magellanic Cloud, allow lucky astronomers to examine the process of star formation in an environment outside the Milky Way. Bright blue newly formed stars are blowing a cavity in the center of the region.

# THE GREAT SPIRAL

The constellation Pegasus, the Flying Horse, is a prominent feature of fall skies in the Northern Hemisphere (and spring skies in the Southern). It is easy to recognize because it contains an almost perfect square, the Square of Pegasus.

The star that marks the top-left corner (bottom-right in the Southern Hemisphere) actually belongs to the linked constellation Andromeda. If you cast your eyes up and slightly to the left (down and to the right in the Southern Hemisphere), you will see a faint, misty patch. It looks like a nebula, but isn't—it's a neighboring galaxy.

The Andromeda Galaxy is nowhere near as close as our other galactic neighbors, the Magellanic Clouds. It lies around 2.9 million light-years away and is the most distant object that is visible to the naked eye.

The Persian astronomer Al-Sufi mentioned the galaxy first in his Book of the Fixed Stars in the year 964 CE. German astronomer Simon Marius spotted it again in a telescope in 1612, saying that it looked like "the light of a candle seen through horn." Charles Messier included it as number 31 in his list of star clusters and nebulae of 1781.

It seems astonishing that we can see the Andromeda Galaxy, M31, at a distance of millions of light-years away. This is because of its enormous size. The galaxy we see with our eyes is about 150,000 light-years across, which is one and a half times the size of the Milky Way. But, observations made in recent years suggest there is material beyond this visible disk and the galaxy is up to six times bigger. It also contains many more stars—infrared observations made in 2006 suggest it has about one trillion. It dominates the Local Group and could, in billions of years, consume most of the other galaxies, including the Milky Way.

M31 is a spiral galaxy, also called the Andromeda Nebula and the Great Spiral. Unfortunately, we don't have a very good view of its spiral arms because we see it from the side. M31 was one of the spiral nebulae investigated in the early twentieth century by Edwin Hubble and other astronomers. It was the first one he proved was extragalactic—beyond our own Galaxy (see page 184).

Later studies of the spiral arms and central bulge by Walter Baade (1944) led to his recognition of the two populations of stars. Population I stars are the relatively young stars of spiral arms, while Population II stars are the older stars of the galactic center.

The Andromeda Galaxy is like our own spiral in another way; it has two close companions, the galaxies M32 and NGC 205. They are both dwarf elliptical galaxies in orbit around the Great Spiral.

ABOVE
**Stars with the blues**
Deep inside the elliptical satellite galaxy M32, the HST has spotted a swarm of hot blue stars, seen here in ultraviolet light. Unusually, these stars seem to be old, at a late stage in their evolution when they are burning helium in their cores.

BELOW
**Andromeda's globulars**
The HST views globular clusters orbiting M31's nucleus with almost the same clarity as round-based telescopes view globulars in our own Galaxy. This Andromeda globular boasts plenty of red giants, as would be expected. The two other bright objects are nearby stars in our Galaxy.

RIGHT
**The spiral giant**
Our close galactic neighbor, the Great Spiral in Andromeda (M31), is the most famous of all the outer galaxies. It appears to have two major spiral arms. Seen here also are the two satellite galaxies—M32 close in and NGC 205 farther out.

BELOW
**Galaxy cluster**
The diverse nature of galaxies within a galaxy cluster is revealed in this HST image of Abell S0740. In the center is a giant elliptical galaxy, which is as massive as 500 billion stars. Other ellipticals and several spirals are also present.

FAR RIGHT
**Ripped apart**
This trio of galaxies is having a galactic tug-of-war which is likely to end as two rather than three galaxies. Together known as Hickson Compact Group 90, two of the three are elliptical galaxies (middle left and lower right); the third (middle right) is a spiral being ripped apart by its neighbors.

RIGHT
**Dance of destruction**
This compact group of galaxies is named Seyfert's Sextet for its discoverer Carl Seyfert. But only four galaxies are physically associated and interacting. The small face-on spiral lies much farther away than the others, and the object at lower right is an elongated tail of stars rather than a galaxy.

# GROUPS, CLUSTERS, AND SUPERCLUSTERS

Galaxies are found throughout the Universe in groups and clusters, and these cluster together to form superclusters. The term "group" generally indicates the presence of only a few tens of galaxies; the Local Group which the Milky Way belongs to consists of only about 40 galaxies. In contrast, clusters can contain up to many thousands of galaxies.

## COMPACT GROUPS
Additionally, there are 100 known small and compact groupings that each contains only a handful of galaxies. These are known as Hickson Compact Groups after the astronomer Paul Hickson who catalogued them in the 1980s. The galaxies within these groups are tightly bound by gravity. Eventually, the galaxies will merge and will no longer be a group.

The best known of these compact groups was the first to be discovered, in 1877. It is a small group of five galaxies and became known as Stephan's Quintet. Today it is also known as Hickson Compact Group 92. The group was imaged by Hubble in 2009 using the recently installed Wide Field Camera 3, and studies have shown that only four of the galaxies are connected. One is actually separate, but it is misleading because it lies on the line of sight between Earth and the rest of the group, which are seven times farther away.

## CLUSTERS
Clusters not only vary in the number of galaxies they contain but also in their shape. The Virgo Cluster, which is the nearest major cluster to us, contains up to about 2,000 galaxies but like the Local Group is irregular in shape. The Coma Cluster contains about one thousand galaxies and is roughly spherical. The size of clusters can also differ but not by as much as you might expect. Clusters take up a roughly similar volume of space. The Local Group with its small number of galaxies occupies a volume that stretches across 10 million light-years. By contrast, the huge population of the Virgo Cluster is densely packed into a volume about 15 million light-years across.

In general, irregular clusters are made up of mainly spiral galaxies, while elliptical galaxies dominate in spherical clusters. Irregular clusters also seem to be much younger than spherical ones. Astronomers think this probably reflects how galaxies evolve. Over time the spirals in irregular clusters tend to collide and merge to form ellipticals, while the cluster itself assumes a more regular shape.

Galactic collisions account for the presence in some large clusters of giant elliptical galaxies. These galaxies are worthy of their name because they can measure as much as

two million light-years across—about a quarter the size of the entire Local Group. The three at the center of the Virgo Cluster are believed to be the result of collisions between spiral galaxies billions of years ago.

## SUPERCLUSTERS

It is gravity, of course, that holds the galaxies together in clusters. And together these form mammoth superclusters. Our Local Group belongs to a supercluster centered on the big Virgo Cluster. Called the Virgo, or Local Supercluster, it occupies a region of space about 100 million light-years across.

Superclusters are the biggest structures in the Universe. So far about 100 have been identified within one billion light-years of us. They extend throughout the Universe in filaments; these flat sheet- and streamer-shaped structures cover distances of more than 100 million light-years. One of the largest is the Great Wall. It measures more than 500 million light-years by about 300 million light-years in area, and is15–20 million light-years thick.

These huge structures are identified as a result of surveys that study galaxy distribution. The first survey was in the late 1980s to early 1990s. The largest to date is the Sloan Digital Sky Survey which started in 2000 and has mapped more than one quarter of the sky. It has produced 3-D maps that plot more than 930,000 galaxies and more than 120,000 quasars.

The largest known supercluster structure is the Great Sloan Wall which was identified in the Sloan survey data in 2003. It is 1.37 billion light-years long and the largest known structure in the Universe. It is located roughly one billion light-years from Earth.

## LARGE-SCALE STRUCTURE

Like stars and galaxies, superclusters are not evenly distributed throughout space. The 3-D maps produced by the galaxy surveys show the superclusters apparently wrapped around empty spherical regions called voids. These voids connect with one another giving the Universe a sponge-like structure. This irregular scattering of galaxies and clusters could well reflect the original "lumpiness" of matter after the Big Bang, as inferred by measurements of cosmic background radiation by the COBE and WMAP spacecraft (see page 172).

LEFT
**Distorted images**
Faraway galaxies appear in this Hubble image as arc-shaped dashes of light around the galaxy cluster Abell 1689. The cluster's gravity has acted as a cosmic lens and bent the light of the far-distant galaxies to produce the multiple distorted images.

RIGHT
**Dark matter**
The same image of Abell 1689 (see left) but this time blue has been added to indicate the presence of dark matter. Hubble cannot see the dark matter directly but its location has been identified by analyzing the cluster's gravitational-lensing effect on the galaxies beyond Abell 1689.

BELOW
**Double ring**
These two images—a distant and a close-up—are the first to show a double Einstein ring in space. The pattern of the two glowing rings, one nestled inside the other, is caused by the bending of light from two distant galaxies by a massive foreground galaxy. All three galaxies lie on the same line of sight.

Looking to the future, if about fifty double Einstein rings could be found scattered over the entire sky, their properties could be used to map the distribution of mass throughout the Universe. It could further be used to map the distribution of dark matter, and also the pressure of dark energy, that unknown form of energy which makes up 72 percent of the Universe and is responsible for the acceleration of the expansion of the outer regions of the Universe.

## THE RIDDLE OF DARK MATTER

When astronomers calculate how strong the gravity of a cluster must be to produce a lensing effect, they come up against a problem. Without exception, the total mass of the galaxies visible in the cluster is not nearly large enough to produce the powerful gravity needed for lensing. In general, a cluster needs 90 percent more mass to produce this effect. Astronomers reckon that this mass must reside in matter that is invisible. They call it dark matter.

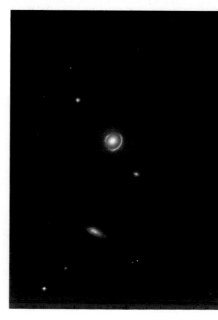

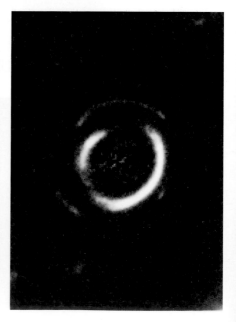

## IN THE HALO

But it is not only in gravitational lenses that dark matter must
exist. The way that spiral galaxies rotate and retain their
structure over time implies that they are constrained by a
gravitational force much greater than their visible matter—their
stars and gas clouds—can account for. Each galaxy must be
embedded within a vast spherical region, or halo, containing
abundant dark matter. The halo, in turn, may be surrounded
by an even more extensive sphere, or corona, of dark matter.
There could be as much as 10 times more dark matter than
visible matter in an average galaxy.

Evidence of dark matter is also provided by galaxy
clusters. The galaxies in clusters travel so fast that, in theory,
they should rapidly escape from one another. The fact that this
doesn't happen suggests that gravity from dark matter must
be holding them back.

RIGHT
**Mapped matter**
The pink regions in
this image of the
galaxy cluster Abell
901/902 identify the
location of dark
matter. The cluster
image is made by an
Earth-based telescope
but the dark matter
was mapped using
Hubble. Astronomers
mapped the dark
matter by analyzing
gravitational lensing
on more than 60,000
galaxies behind Abell
901/902.

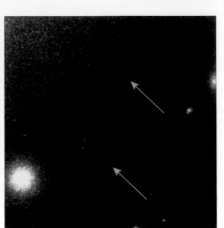

## OF MACHOS …

There is circumstantial evidence that visible matter makes up only 5 percent of the Universe, and about 23 percent is dark matter (the remaining 72 percent is the unknown energy, dark energy). But exactly what is dark matter? Because we can't see it or detect it, we don't know for certain. But it could take a variety of forms.

Ordinary visible matter consists of atoms made up of light-weight electrons and a nucleus of heavier particles, protons, and neutrons. These heavier particles belong to a class of subatomic particles called baryons; so ordinary matter is called baryonic matter.

The dark matter in the halos of galaxies could be invisible baryonic matter, such as the failed stars we call brown dwarfs, dead stars, and black holes. Such invisible objects are called MACHOs (massive compact halo objects).

## … AND WIMPS

Astronomers also believe that the Universe might be flooded with unseen and so-far undetected non-baryonic forms of matter. Physicists have suggested that this matter could consist of WIMPs (weakly interacting massive particles). WIMPs would be difficult to detect because they would barely interact with ordinary matter. Astronomers and physicists are searching for them using specialist telescopes and other detectors.

Other elusive particles that we know exist are neutrinos. Neutrinos produced in the Sun's interior are racing through your body right now. They are elusive because they hardly interact with matter and have no electric charge. Until recently physicists also thought that they had no mass. Current research suggests that neutrinos might have a very small mass—around 1/100,000th the mass of an electron. If this is true, then neutrinos could account for a large proportion of the dark matter in the Universe.

## OPEN OR CLOSED?

The amount of dark matter is crucial in deciding the eventual fate of the Universe. If there is enough matter, visible and invisible, then its collective gravity will be strong enough to halt the Universe's expansion. Otherwise, gravity will not be strong enough and the Universe will continue to expand.

Cosmologists have worked out a critical density for the Universe at which gravity will be just strong enough to stop it from expanding. It works out to be a few hydrogen atoms per cubic yard (or cubic meter). The ratio of the actual density of the Universe to the critical density is designated $\Omega$ (omega).

There seem to be three main possible fates for the Universe. If $\Omega$ equals 1, then the Universe will eventually stop expanding, but only after an infinite amount of time, and it will exist forever. This concept is called the flat Universe.

If $\Omega$ is more than 1, then the Universe will eventually slow down and stop expanding. Gravity will then rein in the galaxies and the Universe will contract. Over time, everything in the Universe will come together in a Big Crunch, a reversal of the Big Bang. This concept is called the closed Universe.

If $\Omega$ is less than 1, the expansion of the Universe that we witness today will continue forever. All the stars, all the galaxies will fade and die; all the black holes will evaporate away. Unimaginably cold, unimaginably vast, the Universe will end up as an ocean of subatomic particles with no energy or ability to interact. This concept is called the open Universe.

So is the Universe flat, closed, or open? Will it exist forever, will it end in a Big Crunch, or will it fade away? The jury is still out, but the signs indicate that ours is an open Universe. And recent observations suggest that the rate of expansion is actually increasing. This has led to the suggestion that a mysterious force called dark energy must be stretching the Universe.

# 5 | Solar Systems

**Other stars have planetary systems too**

ABOVE
**Stellar beginnings**
Dark dust clouds and hot young stars make up this star-forming region, N11B, in the Large Magellanic Cloud, a nearby galaxy. The blue and white stars to the left are among the most massive stars known in the Universe.

INSET LEFT
**Comet Tempel 1**
This HST image of Comet Tempel 1 was taken just days before the Deep Impact spacecraft made its rendezvous with the comet. The comet's inner coma of dust and gas surrounds the cometary snowball nucleus.

INSET RIGHT
**Asteroid Ida**
The probe Galileo sent back this image of the asteroid Ida while it was on its way to Jupiter in 1993. Ida measures about 35 miles (55 km) long, and incredibly, has a minuscule moon orbiting around it, named Dactyl.

# PLANETS IN THE MAKING

Vast, dark tenuous masses of gas and dust called giant molecular clouds are found throughout interstellar space. They are the stuff that stars are made of; are born from. Five billion years ago, a giant molecular cloud occupied our little corner of the Universe. It was from this that the Sun and the planets were born.

Part of the cloud began to collapse under gravity and rotate faster. As the collapse continued and rotation became faster still, the cloud flattened out to form an embryonic solar system that astronomers call the solar nebula. The denser center of this rotating mass began to heat up and glow, becoming the proto-Sun.

Matter continued to feed into the collapsing proto-Sun, which in turn continued to heat up. Eventually, soaring temperatures and pressures triggered nuclear reactions within its core, and it began to shine as a new star. This fledgling Sun then spent roughly 10 million years calming down, until it reached equilibrium between the outward pressure of radiation and the inward force of gravity. It became a stable main-sequence star.

Meanwhile, the surrounding matter had flattened itself into a disk, and it was from this disk that the planets and other members of the solar system eventually emerged.

Scenarios similar to the birth of our solar system are being enacted throughout interstellar space, giving rise to planetary systems around other stars. The first evidence of possible external solar systems came in 1983, when the Infrared Astronomy Satellite (IRAS) detected dusty disks around the stars Beta Pictoris and Vega. The HST has since spotted dusty disks around a host of stars—dozens in the Orion Nebula alone. They are termed proplyds, for protoplanetary disks, because they are almost certainly planets in the making.

Within our own solar system, how did the planets we know today form out of the primeval dusty disk that girdled the youthful Sun? Surely the planets of other stars would evolve through a similar process.

## INNER PLANETS

In the flattening disk of matter that formed around the proto-Sun, particles began to clump together. Eventually, these clumps grew into masses several thousand yards (or meters) across, known as planetesimals. By this time, these great chunks were in orbit around the newborn Sun.

The temperature of the disk of rotating matter varied widely, from a high of around 3,500 degrees Fahrenheit (2,000°C) closest to the Sun to a low of only about −360 degrees Fahrenheit (−220°C) at the outer edge. This variation in temperature dictated the phase of the material; solid or gaseous.

In the hot, inner region, metals and rock-forming silicate minerals (which have high melting points) collected. Planetesimals with this composition gradually merged by collision to form larger and larger objects, which eventually became the rocky, terrestrial planets—Mercury, Venus, Earth, and Mars.

It took roughly 10 million years for these planets to reach more or less their present size, and about another 100 million years for them to sweep up any remaining planetesimals. By this time, powerful blasts of particles in the solar wind had swept the inner solar system clear of gas and volatile substances like water, ammonia, and methane.

RIGHT
**Stars with planets**
The top image is a view towards the center of our galaxy in the constellation Sagittarius and contains approximately 150,000 stars. The HST monitored the stars for possible planets passing in front of them. The blue circles identify stars orbited by planets every few days. One of these stars is shown in the detailed picture, below. The planet orbiting SWEEPS-04 (Sagittarius Window Eclipsing Extrasolar Planet Search) is estimated to be less than 3.8 Jupiter masses.

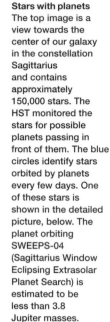

RIGHT
**Possible planet?**
A dust disk surrounds the star Beta Pictoris whose light has been blocked out in this HST image. The 2nd less-extensive disk tilted to the 1st could indicate the presence of a planet. The planet may have formed the 2nd disk by sweeping up material from the main disk.

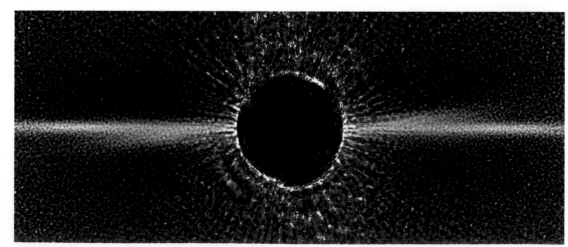

SWEEPS -04

**Orion proplyds**
The HST has imaged more than a hundred protoplanetary disks, or proplyds, in the vast star-forming region that is the Orion Nebula. Clearly, they are a common by-product of star formation.

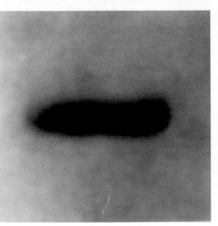

## OUTER PLANETS

Planets were also forming in the outer part of the circumsolar disk, where it was cooler. Hundreds of millions of miles away from the Sun, bodies were growing that would become Jupiter and Saturn. They were growing huge by attracting gas, primarily hydrogen and helium, from their surroundings.

Farther out still, the bodies that would become Uranus and Neptune were created when the volatiles, such as water and ammonia, condensed on small, rocky cores. They accumulated relatively little hydrogen and helium, since Jupiter and Saturn had mopped up most of those gases.

Beyond Neptune, ice was the predominant matter. Relatively small worlds of ice and rock formed, of which Pluto was one of the largest. Beyond these worlds and out to the edge of the solar system, matter remained as icy planetesimals just a few miles across. The bulk of these icy bodies remain today. We occasionally see them when they journey in toward the Sun and start to shine—as comets.

## EXTRASOLAR PLANETS

It would be statistically impossible for there not to be other planetary systems like our own. The process of planetary formation in the solar system can't be unique, for this would suppose that the birth of the Sun was a freak occurrence, and we know that this was not the case. And there are all those protoplanetary disks that the HST has spotted that are planets in the making. The question is, how do we prove that there are extrasolar planets—planets beyond our solar system? The answer is: with great difficulty, because planets are tiny compared to stars and don't give off light of their own. Nevertheless, astronomers can still detect them.

The first extrasolar planets were discovered in 1991, not around an ordinary star, but around a pulsar—the remnant of a star that blew itself apart. It was not until 1995 that the first planet was found around an ordinary star, 51 Pegasi. Since then more than 500 extrasolar planets have been discovered. They seem to be roughly the same order of size as Jupiter, but usually orbit much closer to their parent star.

Astronomers discover these planets by indirect observation. The main technique they use is to look for stars with a characteristic "wobble." A wobble, or slight motion to and fro, tends to indicate that one or more planets is orbiting the star and affecting it gravitationally. Astronomers detect wobbling stars by examining the spectrum of their light. Their motion causes the dark lines in the spectrum of their light to shift, first toward the blue, then toward the red.

If planets are as common as we think, what are the chances of there being planets like Earth? Good. And could such planets harbor life, intelligent or otherwise? Probably. It is with this in mind that some astronomers are involved in SETI, the search for extraterrestrial intelligence. The giant radio telescope at Arecibo in Puerto Rico is at the forefront of current SETI research (see page 168).

BELOW
**Fomalhaut's planet**
The star Fomalhaut is the small white dot in the center of the huge red dust disk. Its brightness has been greatly reduced in this Hubble image to show the dust and its planet. The inset detail shows the planet in 2004 and in 2006. The planet is a billion times fainter than its star, takes 872 years to orbit it, and is 10.7 billion miles away from it.

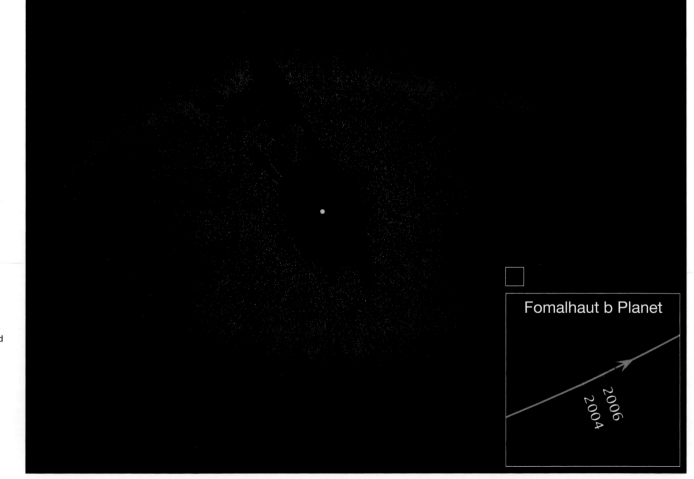

LEFT
**Interstellar frisbees**
Looking rather like frisbees, these are some of the protoplanetary disks that have been found in the Orion Nebula. They range in size from about 2 to 17 times the size of our own solar system.

Fomalhaut b Planet

2006
2004

# FAMILY OF THE SUN

It was the astrologers of early civilizations in the Middle East who laid the foundations of astronomy. They became familiar with the heavens, but had absolutely no idea of what the Universe was really like. They did, however, all agree that the Earth was its center.

The belief that the Earth was flat had, by Aristotle's time (the fourth-century BCE), given way to the idea of a round or spherical Earth. Aristotle pointed out that the Earth threw a curved shadow on the Moon. Furthermore, the sphere was the perfect shape and therefore appropriate for a body that was the center of the Universe. The dissenting voice of Aristarchus of Samos, a next-generation philosopher who argued that the Earth might orbit the Sun, fell on deaf ears.

Much later, around 150 CE, the Alexandrian astronomer Ptolemy elaborated the classical Earth-centered view of the Universe, which became known as the Ptolemaic system. According to this system, all the heavenly bodies—Sun, Moon, planets—circled around a central Earth. The stars were fixed to the inside of an all-enveloping dark orb, the celestial sphere, which rotated around the Earth once a day.

This concept held sway throughout classical times and the Dark Ages that followed, when astronomy, like so many other branches of learning, suffered almost terminal neglect.

ABOVE
**The Copernican system**
This old print (1708) depicts Copernicus's Sun-centered Universe.

BELOW

**The solar system**
Orbits of the eight planets and the dwarf planet, Pluto, in our solar system. The planets are widely scattered, traveling within a region of space 8 billion miles (14 billion km) across. The real edge of the solar system, far beyond Pluto's orbit, is marked by an enormous sphere of cometary bodies called the Oort Cloud.

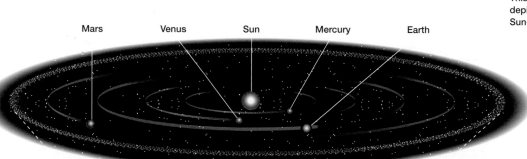

Mars     Venus     Sun     Mercury     Earth

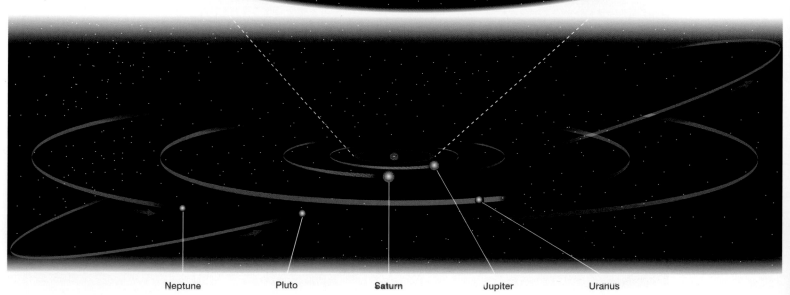

Neptune     Pluto     Saturn     Jupiter     Uranus

Only in the Arab world did the art and craft of astronomy continue to advance. This explains why many stars have Arab names, such as Betelgeuse (in Orion), Algol (in Perseus), and the descriptive Zubenelgenubi and Zubeneschamali (Southern Claw and Northern Claw) in Libra. The names of the last two reflect that they were once considered part of Scorpius (the Scorpion).

## THE COPERNICAN REVOLUTION

In the fourteenth and fifteenth centuries came the rebirth of learning we call the Renaissance. People began to question age-old beliefs, and science (including astronomy) began to advance once more. In 1543, a Polish religious cleric with a passion for astronomy upset the old order and antagonized the Church by putting forward the concept of a Sun-centered, or solar system. His name was Nicolaus Copernicus.

By carefully examining his own and others' observations, Copernicus became convinced that the often bizarre motion of the planets—that they sometimes back-track through the heavens—could better be explained if they and the Earth itself circled the Sun. It took nearly a century for other astronomers, and the Church, to accept the Copernican solar system, which in effect marked the dawn of modern astronomy.

## OUR SOLAR SYSTEM

What is this solar system of ours like? The main bodies are the eight planets and their satellites, or moons. But there are also many other smaller bodies, including dwarf planets, asteroids, Kuiper Belt objects, and comets.

The planets circle round the Sun at different distances. They don't exactly "circle" the Sun, but rather travel around it in elliptical (oval-shaped) orbits. All eight planets orbit around the Sun in much the same plane (flat sheet) in space. They also all travel in the same direction. If you could view the solar system from a point in space above the North Pole, the planets would orbit the Sun counter clockwise. The planets also have another motion: they spin around in space. Earth spins around once in just 24 hours, but the other planets have different spin times, from under 10 hours (Jupiter) to 243 days (Venus).

There are the four terrestrial planets made of rock (Mercury, Venus, Earth, Mars) and four giant gassy planets (Jupiter, Saturn, Uranus, Neptune). The five dwarf planets are Ceres, the largest asteroid; Eris, the largest Kuiper Belt object; and the three Kuiper Belt objects Makemake, Haumea, and Pluto, which was classed as a planet until 2006.

What keeps the planets circling round the Sun? It is the Sun's enormous gravity—enormous because the Sun has such an enormous mass—750 times more than the mass of all the other bodies in the solar system combined. The Sun thus has much better credentials than the Earth for being the center of our corner of the Universe.

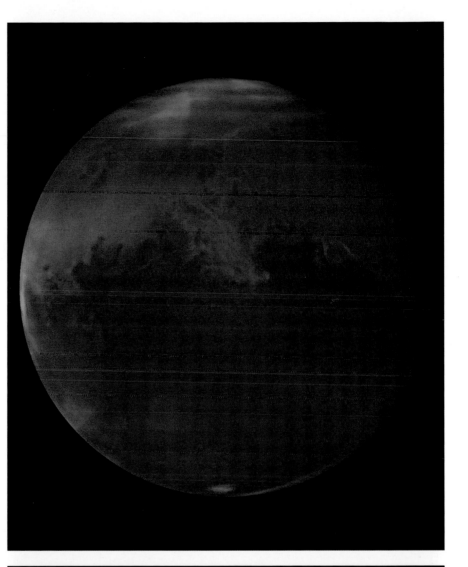

TOP
**Dusty Mars**
The HST captured Mars on 28 October 2005 a day before its closest approach to Earth.

ABOVE
**Serene Saturn**
HST images use four filters to render the colors as seen by the eye through a telescope.

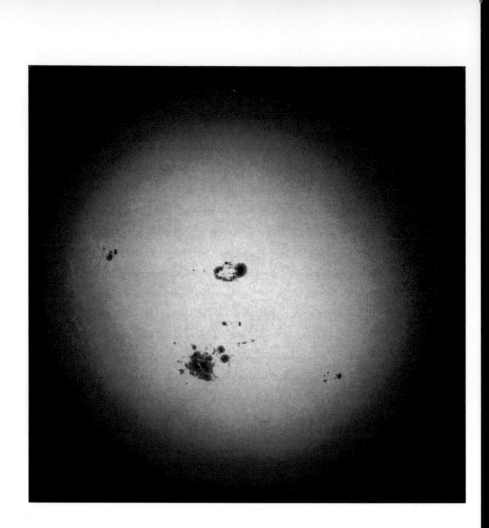

# NEIGHBORHOOD STAR

The star at the center of our own solar system, the Sun, is special to us. But in the Universe as a whole, it is very ordinary. It is one of at least 500 billion stars in our Galaxy alone. It may seem big to us, but as stars go, it is a dwarf. The biggest supergiant stars such as Betelgeuse in Orion are hundreds of times bigger in diameter. However, for astronomers, the Sun is of extraordinary importance because it is the only star they can study from close quarters. The Sun is a mere 93 million miles (150,000,000 km) away, while the other stars are light-years from us.

Like other stars, the Sun is made up mainly of hydrogen and helium, the two most common elements in the Universe. There are also traces of as many as 70 other elements. All these elements are present as plasma, a form of matter in which the atoms exist as ions (charged particles) in a sea of electrons. Matter takes this form at the kinds of temperatures that exist in the Sun and stars.

## SOLAR ENERGY

The temperature of the bright surface of the Sun, the photosphere, is relatively low, about 10,000 degrees Fahrenheit (5,500°C). As you go deeper into the Sun's interior, temperatures rocket. In the core (center), they reach 27 million degrees Fahrenheit (15,000,000°C).

At such astronomical temperatures and pressures, nuclear reactions take place, which produce the energy that keeps the Sun shining. It has been shining steadily for nearly five billion years, and should continue for as long again.

In these nuclear reactions the nuclei (centers) of atoms of hydrogen fuse (join) to form nuclei of helium atoms. In the process small amounts of mass seem to be destroyed, but in fact, they are transformed into energy. Albert Einstein quantified this mass-energy transformation in that most famous of mathematical equations: $E=mc^2$, where $E$ is the energy released when a mass $m$ is transformed, and $c$ is the velocity of light. Since $c$ is a huge value—186,000 miles (300,000 km) per second—the energy released when even a small amount of mass is converted is incredibly large. In practice, six million tons of hydrogen are converted into energy every second.

RIGHT
**The Sun's crown**
In a total solar eclipse, the Moon moves in front of the glaring surface of the Sun and blots out its light. It is then that we can see the Sun's pearly white outer atmosphere, the corona (crown).

FAR RIGHT
**Lunar transit**
When the STEREO spacecraft pointed its cameras at the Sun on February 25th 2007 they got an extraordinary view. The black disk of the Moon was seen transiting the Sun.

FAR LEFT
**Spotty Sun**
Sunspots are dark solar regions around 2,000 degrees cooler than the normal surface. This spot group (below center) was seen in October 2003, and was one of the biggest of the last solar cycle. The surface area was about fifteen times greater than the width of Earth.

LEFT
**Solar Prominence**
This gigantic loop of magnetically constrained gaseous plasma is a solar prominence. Captured by the STEREO spacecraft on September 30th 2010, it is short lived and unstable. On disruption the gas will be blown away from the Sun and the solar system.

## THE SUN'S RAYS

The energy produced in the Sun's nuclear furnace travels to the surface first as radiation, then on convection currents of rising gas. It pours out from the photosphere as visible light and other invisible radiation, such as gamma rays, X-rays, ultraviolet rays, infrared (heat) waves, microwaves, and radio waves. These are different kinds of electromagnetic waves—minute electrical and magnetic disturbances in space. They differ in their wavelengths, with gamma rays being the shortest and radio waves the longest.

The HST was designed to study the light and ultraviolet and near-infrared radiation from the distant stars, but not from the Sun. The intensity of the Sun's radiation would effectively burn out all the sensors. Wisely, HST designers equipped the telescope with a protective sunshade.

## ESSENTIAL SUN

| | |
|---|---|
| Diameter at equator: | 865,000 miles (1,392,000 km) |
| Average distance from Earth: | 93 million miles (149,600,000 km) |
| Spins on axis in: | 25 days at equator |
| Mass: | 333,000 times Earth's mass |
| Surface gravity: | 28 times Earth's gravity |
| Surface temperature: | 10,000°F (5,500°C) |
| Core temperature: | 27 million°F (15,000,000°C) |

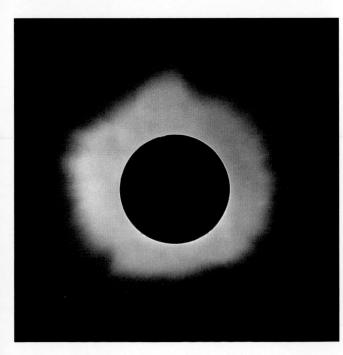

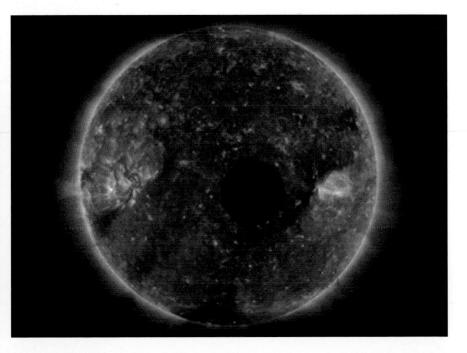

# THE SILVERY MOON

One of the oddities of the heavens is that, from Earth, the Sun and the Moon appear to be almost identical in size. Nothing could be further from the truth. With a diameter of 2,160 miles (3,476 km), the Moon is actually only 1/400th the size of the Sun. But, because the Moon is 400 times nearer to us than the Sun, the two bodies appear to be the same size.

Of course, the Moon is quite a different body from the Sun, being made of solid rock rather than incandescent gas. And whereas the Sun shines in its own right, the Moon shines only because it reflects sunlight, though not particularly well.

The Moon is Earth's only natural satellite, orbiting our planet about every four weeks. During this time, because of the geometry of the Sun, Earth, and Moon in space, we see more or less of the Moon lit up by the Sun. This makes the Moon appear to change shape in the sky—from a thin crescent, to full circle, and back again. These changing shapes, or phases, of the Moon mark one of the great rhythms of nature.

## THE LUNAR SURFACE

With our eyes, we can make out two general features on the Moon—dark and light areas. The dark areas were once thought to be seas and were named maria (Latin for "seas," singular "mare"). But in reality they are dusty plains. The lighter areas are highland regions, thought to be remnants of the Moon's original crust.

Many of the Moon's mare regions are circular, and this shape suggests their origin. Early in its history, the Moon was pounded by asteroids, and the largest ones gouged out huge basins from the surface. Subsequently, volcanic activity filled the basins with lava, resulting in the dark, flat landscapes that typify the maria.

The fact that the maria were formed more recently than the Moon's highlands is evidenced by the number of craters. On the maria, craters are few and far between, whereas in the highlands they crowd one another. The largest of these craters, such as Bailly and Clavius, measure over 150 miles (250 km) across.

RIGHT
**Moon rock**
A fine-grained volcanic rock brought back from the Moon by Apollo astronauts. Similar to terrestrial rocks called basalt, it is riddled with holes where gas bubbled out of cooling lava.

RIGHT
**Taurus-Littrow**
Harrison Schmitt examines a huge split boulder near the Apollo 17 landing site at Taurus-Littrow, on the edge of the Sea of Serenity. As ever, the lunar landscape is hauntingly beautiful.

Thanks to close-up investigation by probes and the exciting expeditions by the Apollo astronauts, we know exactly what the Moon is made of. The surface is covered with a kind of soil that crumbles easily, known as regolith. All the rocks are volcanic in origin. The main types are basalt, much like Earth basalts, and breccia, made up of chips of preexisting rocks cemented together.

Some of the moons that orbit other planets resemble our Moon superficially. Many are heavily cratered from heavy bombardment from outer space. But they have a quite different composition. Most of these moons are made up of varying mixtures of ice and rock—Jupiter's moon, Io, is one of the exceptions. It is a yellow–orange volcanic world.

Our Moon is so close that the HST would have to take 130 separate images to cover the Moon's entire face. The telescope is more often used to target moons of distant planets. In May 2005 it discovered two new moons around Pluto.

## ESSENTIAL MOON

| | |
|---|---|
| Diameter at equator: | 2,160 miles (3,476 km) |
| Average distance from Earth: | 239,000 miles (384,000 km) |
| Spins on axis in: | 27 days, 7 hours |
| Circles Earth in: | 27 days, 7 hours |
| Goes through its phases in: | 29 days, 12 hours |
| Mass: | 1/81 Earth's mass |
| Surface gravity: | 1/6 Earth's gravity |

# BROOM STARS

Of all the objects that light up the night sky, none are as spectacular or as intriguing as comets. Today, when thousands of astronomers routinely scan the skies, there is little chance of a comet sneaking up on us undetected, as happened in the past, although only the brightest are detected.

In past times, comets seemed to appear suddenly out of nowhere, with their glowing heads and long tails fanning out behind them. To superstitious peoples, the appearance of a comet was a bad omen, a portent of evil, a harbinger of drought or disease, death, and destruction. Over 2,500 years ago the Chinese seemed to take a particular interest in comets, which they called broom stars. Ancient Chinese scribes wrote that comets were vile. King Harold of England would probably have agreed, for he was shot in the eye when a comet appeared at the Battle of Hastings in 1066.

## A COMET CALLED HALLEY

The comet of 1066 was actually making one of its regular appearances in Earth's skies. English Astronomer Royal Edmond Halley was first to recognize this some 600 years later, and so the comet was named after him. Halley's Comet was last seen in Earth's skies in 1986 and will return in about 2061. Thousands of the world's astronomers and a flotilla of spacecraft studied the 1986 appearance of the comet in detail, though the HST had yet to be launched at that time.

Visually, however, the 1986 visit of Halley's Comet was a disappointment. It was faint and difficult to spot among the stars, and observers needed a telescope to see its tail. Comet-watchers had to wait a decade before seeing the first of a pair of spectacular comets with the naked eye. The first was Hyakutake in 1996, the second Hale-Bopp a year later. Hale-Bopp in particular was outstanding—it was one of the great comets of the twentieth century. It hung in night skies for months, outshining all but the brightest stars. In early 2007, Comet McNaught was also easy to see with the naked eye.

BELOW
**Hyakutake Telescope and HST (inset) images of the 1996 comet Hyakutake. Easily visible to the naked eye, it grew a long tail. Hubble images of the comet focused on the region around the nucleus and showed bits that had broken off.**

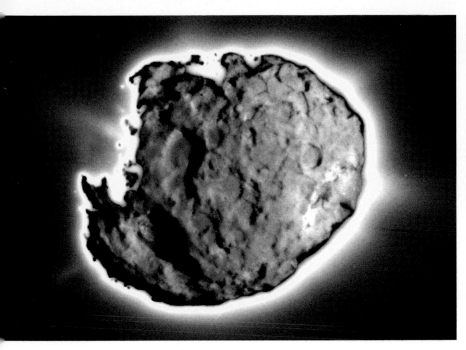

## DIRTY SNOWBALLS

Astronomers study comets closely, not just because they are so spectacular, but also because they are perhaps the most primitive bodies in the solar system. They seem to originate in a vast reservoir of comets at the outer fringes of the solar system, called the Oort Cloud. This reservoir was created when the planets were being born, and the bodies it contains have remained virtually unchanged since then. They therefore hold the key to understanding the origin and evolution of the solar system.

What are these primitive bodies like? They are made up mainly of a mixture of ice and dust and are often described as dirty snowballs. Far away, in the dark, cold depths of the outer solar system, they remain in deep-freeze. They become visible as comets only when they travel toward the Sun and begin to absorb some of the Sun's warmth.

As they warm up, the icy surface starts to evaporate, releasing clouds of vapor or gas. The gas spurts out in jets, carrying dust with it. The gas and dust form a cloud around the solid part, or nucleus, of the comet. The cloud reflects sunlight, and the comet becomes visible. The cloud can measure up to hundreds of thousands of miles across, but the nucleus itself is usually only a few miles wide.

The pressure of radiation from the Sun forces dust away from the glowing head of the comet, or coma, and shapes it into a fan-shaped, yellowish tail. The gas particles fluoresce due to absorbing sunlight, and magnetic effects force them into a tail called the ion tail (ions are electronically charged particles). In the biggest comets, the tails can stretch for up to 100 million miles (160,000,000 km).

The HST was used to study comets shortly after it was launched and produced spectacular images of Shoemaker-Levy 9 (see overleaf). It also set its sights on Hyakutake and Hale-Bopp. In April 2006 the HST provided astronomers with extraordinary images of Schwassman-Wachmann 3. The comet was disintegrating before their eyes.

"I think I've found a squashed comet!" exclaimed U.S. astronomer Carolyn Shoemaker on March 24th 1993. She was examining a photographic plate from a recent exposure during a standard search for asteroids she regularly conducted at Palomar Observatory in California, with her husband Gene (Eugene) Shoemaker and their colleague David Levy.

In turn, the other two checked the film, they saw a strange apparition with a bar-shaped nucleus, surrounded by a series of little tails and two "wings" of light. None of them had ever seen anything like it, and they were the experts. Together they had already discovered eight comets.

Levy called his friend Jim Scotti at Kitt Peak Observatory in Arizona, who directed his Spacewatch telescope to the position of Carolyn's "squashed comet." Scotti's telescope revealed that it was actually a string of individual comets traveling close together, each with little tails. They were almost certainly pieces of a larger body that had broken up.

The comet-watch trio reported their find, which became known officially as Periodic Comet Shoemaker-Levy 9 (SL9).

In the weeks that followed, astronomers around the world trained their telescopes on SL9. Tracing its path through the heavens revealed that the comet was in orbit around Jupiter. Calculations showed that it must have been the giant planet's powerful gravity that had ripped the larger body apart around July 8th 1993, when it had passed within 13,000 miles (21,000 km) of Jupiter's cloud tops.

Soon, images from a bevy of telescopes, including the HST, showed 21 cometary fragments stretched out in formation—a celestial string of pearls. They seemed to be about two-thirds of a mile (1 km) across.

On July 16th 1993, SL9 reached apojove, the most distant point in its orbit around Jupiter, and began to home in on the planet. But calculations showed that this time SL9 would not skim over Jupiter's cloud tops. It would surrender completely to Jovian gravity and

ABOVE
**String of pearls**
In May 1994, two months before impact, the 21 icy fragments of Shoemaker-Levy 9 (SL9) are strung out across more than 700,000 miles (1,100,000 km). The HST recorded this image in red light.

ABOVE INSET
**Fireballs**
The successive impacts of the comet fragments created huge fireballs in Jupiter's atmosphere, which telescopes on Earth could picture in infrared light.

impact the planet. Astronomers estimated that the first fragment would hit the surface of Jupiter in July 1994.

Sure enough, on July 16th 1994, fragment A plunged into the Jovian atmosphere, creating a fireball that rose 600 miles (1,000 km) above the cloud tops. One by one, over the next seven days, the fragments bombarded the giant planet, creating fireballs and staining the atmosphere with dark clouds.

Most spectacular was the impact of fragment G on July 18th 1994, which sent a fireball leaping 2,000 miles (3,000 km) beyond the cloud tops. Estimated to have released the energy equivalent to six million megatons of TNT, it was three billion times more powerful than the atomic bomb that destroyed the Japanese city of Hiroshima during World War II.

This was the first time in history that astronomers had witnessed such a cosmic collision, and they were ecstatic. But it brought home to humanity just how vulnerable planets, including Earth, must be to bombardment from outer space.

LEFT
**G impact**
This mosaic of images shows the development of the impact site of SL9's G fragment. Just visible in the first image (lower left) is a tiny plume, marking the moment of impact. The second image shows the impact site 90 minutes later. The third shows the site three days later, along with the site of the L fragment impact. The fourth shows the same region two days later still.

# SHOOTING STARS AND ASTEROIDS

The Earth is under constant bombardment from outer space, as a night of stargazing should demonstrate. From time to time you may see bright streaks in various parts of the sky. It looks as if stars are shooting across the sky or falling to Earth.

However, the shooting or falling stars you see are meteors—bright streaks made by specks of matter burning up in the atmosphere. Interplanetary space is full of specks and larger lumps of rock and metal, particles we call meteoroids.

Some meteoroids have been chipped off of asteroids or even other planets. Others come from comets, which shed material into space on their passage around the Sun. The intense displays of meteors at certain times of the year, known as meteor showers, are associated with the orbits of particular comets.

The meteoroid particles that cause meteors—usually barely larger than sand grains—burn up to ash, which slowly falls and settles on the ground. Occasionally, bigger meteoroid lumps plummet through the atmosphere, to the accompaniment of more spectacular *son et lumièrè* (sound and light). The "lumièrè" can be a multicolored fireball, the "son" a muffled sonic boom. Some of the meteoroid survives and falls to Earth, and becomes what we call a meteorite.

## ERRANT ASTEROIDS

If they are big enough, meteorites can blast out mile-wide craters and pose a real threat to life and limb. But asteroids present a much greater threat to humanity.

Asteroids are a group of much bigger lumps of rock and metal, found mainly in a broad band, or belt, between the orbits of Mars and Jupiter. Even the biggest, Ceres, is less than 600 miles (1,000 km) wide, and only about 100 of the 260,000 catalogued so far are bigger than 125 miles (200 km) across. Most are irregular in shape, but the largest are near spherical.

The asteroids appear to be the remains of a failed planet, planetesimals that could not unite because of gravitational disturbances, probably principally from nearby Jupiter.

However, not all asteroids orbit within the asteroid belt. Many wander way beyond it, toward Saturn. Others stray in the other direction and wander among the inner planets. Some pursue orbits that take them uncomfortably close to Earth. These are classed as NEOs, or near-Earth objects.

Historically, Earth has almost certainly been hit by asteroids hundreds of times. The impact of an asteroid about 10 miles (16 km) wide is thought to have caused a mass extinction of species, including the dinosaurs, 65 million years ago.

BELOW
**Blue streak**
A curved streak appears on this image of the Milky Way in Centaurus, which looks toward the center of the Galaxy some 25,000 light-years away. It is a trail made by a tiny asteroid a few light-minutes away.

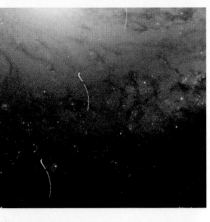

LEFT
**Interlopers**
Many other asteroid trails have been detected in Hubble images. Most of the bodies are too small to be detected from Earth, and their orbits are unknown.

The chances are that some time, sooner or later, a NEO will target Earth. The hope is that we shall have sufficient warning to try to do something about it. Several organizations around the world now actively look for NEOs. One is the University of Arizona's Spacewatch, with facilities at Kitt Peak Observatory. Discoveries made by such organizations regularly spark off media hype that "the end of the world is nigh."

## CLOSE CALLS

Convincing evidence that Armageddon by asteroids is not just science fiction came within the span of a few weeks in the summer of 2002. First, a NEO (2002 MN), 300 feet (91 m) in diameter, was spotted close to Earth, but receding into space. That was the good news. The bad news was that, three days before, it had missed Earth by less than 80,000 miles (120,000 km).

Not long after, another asteroid (2002 NT7) was spotted much farther away, but it appeared to be making a beeline for Earth, to impact on February 1st 2019. However, subsequent refinement of its orbit suggested that it posed no immediate threat.

Then, a few weeks later, NEO 2002 NY40 flashed past Earth with a close approach of about 300,000 miles (500,000 km). Seen with powerful binoculars, it provided visible evidence that the threat from space is chillingly real for us on Earth.

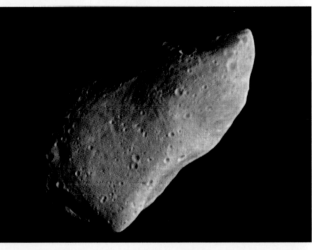

LEFT
**Asteroid Gaspra**
This image of Gaspra was taken by the Galileo spacecraft as it flew by within 3,300 miles, on its way to Jupiter. Rocky Gaspra is 12 miles long, by about 7 miles wide and deep. Its irregular shape indicates that it was once part of a much larger body. It is too small for gravity to pull it into a sphere.

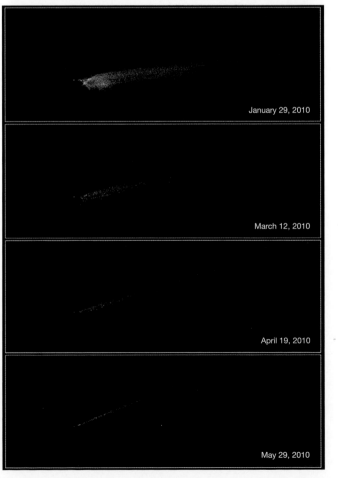

January 29, 2010

March 12, 2010

April 19, 2010

May 29, 2010

LEFT
**Asteroid collision**
This series of Hubble pictures shows the aftermath of a collision between asteroids. The first, taken on January 29th 2010, shows a cloud of rocky, dusty particles ejected from an 80-foot (24.4-meter) wide crater blasted into the surface of an asteroid by the impact of a small asteroid just a few feet (or meters) across. The other pictures were taken on March 12th, April 19th, and May 29th. The debris cloud is getting bigger but appears to shrink simply because Earth and Hubble are getting farther and farther away.

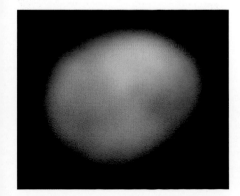

LEFT
**Asteroid Vesta**
The Hubble WFPC2 was used to take this image of Vesta in May 2007. Vesta makes up about 10 percent of the total mass of the asteroid belt, and after dwarf planet Ceres, is the second largest Main Belt asteroid. Vesta measures about 300 miles across.

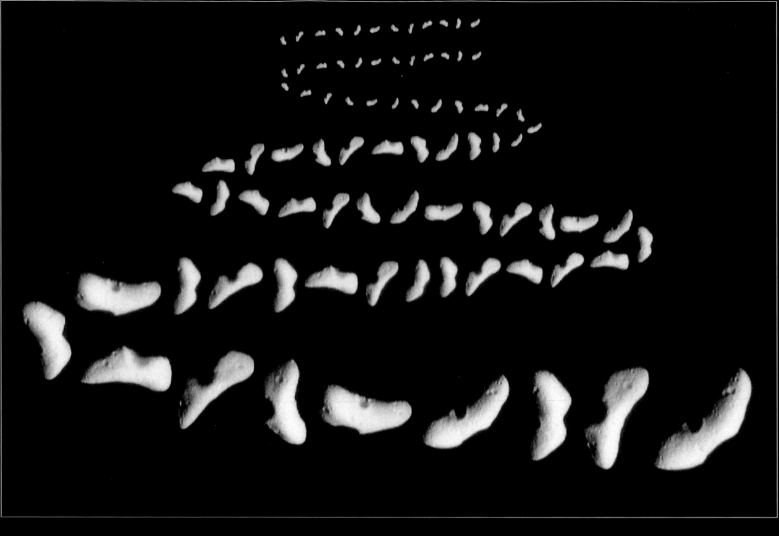

## PROFILING:
# EROS

Dutch astronomer Carl Gustav Witt discovered asteroid number 433 in 1898. It was called Eros for the son of the Greek love goddess Aphrodite. (In Roman mythology, he was called Cupid and was the son of Venus.)

Eros was the first asteroid found to travel mainly inside the orbit of Mars. It is one of the largest NEOs but fortunately doesn't come much closer to Earth than about 14 million miles (22 million km). Historically, astronomers used observations of Eros early last century to calculate the first really accurate value for the astronomical unit—the distance between the Earth and the Sun.

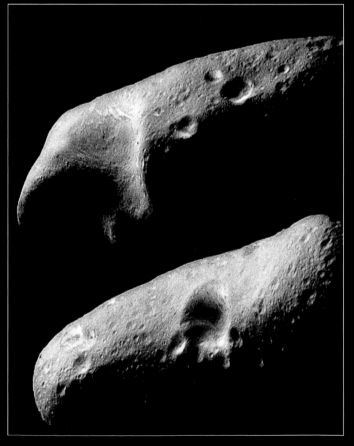

ABOVE
**The approach**
A montage of images taken over three weeks in January and February 2000 as NEAR-Shoemaker homed in on asteroid 433–Eros.

LEFT
**Two hemispheres**
Two opposite hemispheres of Eros are shown in this pair of images, taken from 220 miles (350 km). Left of the top image is the saddle-like feature Himeros. At the center of the lower image is the large crater Psyche.

Even from Earth, astronomers can figure out that Eros is an elongated body that rotates every five hours or so—they can tell this from the way its brightness varies. And they find that it circles round the Sun every 642 days. Eros is one of the most common kinds of asteroids, an S-type; the S standing for silicates, the main mineral ingredients of such asteroids.

We now have a wealth of other data about Eros, for it has been the focus of one of the most incredible feats of interplanetary navigation ever. A spacecraft has not only flown past it, but it has gone into orbit around it and then landed. This spacecraft was NEAR (Near Earth Asteroid Rendezvous Mission).

NEAR was launched in February 1996 and took four years to reach its target. Its trajectory took it several times around the Sun and also included a gravity boost from Earth in January 1998. After a failed attempt at orbiting Eros in December 1998, NEAR achieved success on February 14th 2000. It was appropriate that on Earth it was the day for lovers—Valentine's Day. Once in orbit around Eros, the spacecraft was renamed NEAR-Shoemaker, for the U.S. planetary geologist and comet-hunter Eugene Shoemaker (see page 138).

NEAR-Shoemaker spent almost exactly a year orbiting Eros at a height of about 30 miles (50 km). It took thousands of close-up images of the asteroid, which proved to be elongated as expected and measured about 21 miles (33 km) long and 8 miles (13 km) across. As well as having a camera for imaging, the spacecraft carried spectrometers to measure the asteroid's composition and a laser ranger-finder to map its topography.

Like other asteroids that have been imaged, Eros is pockmarked with craters, though it also has smooth regions as well. One of the largest craters, Psyche, is more than 3 miles (5 km) across. Another prominent feature, Himeros, has a saddle shape. Much of Eros is criss-crossed with linear features that were probably created when the tiny world shuddered under impacts with other asteroids. It seems to be covered in a kind of dusty soil, or regolith, presumably produced when the surface rocks are pulverized by meteorite impacts.

During its year-long orbit of Eros, NEAR-Shoemaker greatly exceeded expectations, and its two-billion-mile (3,200,000,000 km) mission ended in triumph when it landed on the asteroid's surface on February 12th 2001—something it hadn't been designed to do. After four-and-a-half hours of de-orbiting and braking manoeuvers, it touched down at 4 mph (6 km/h).

As the spacecraft was descending, it snapped close-up pictures of the surface from altitudes down to 400 feet (120 m), imaging details as small as half-an-inch (1 cm) across. Mission controllers maintained communications with NEAR-Shoemaker for more than two weeks afterward. Said mission director Robert Farquhar in a masterly understatement: "Things couldn't have worked out better."

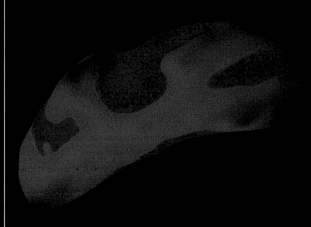

LEFT
**Bizarre grades**
The elongated shape, density, and spin of Eros combine to create curious gravitational topography—uphill and downhill. In this computer model of the asteroid, for example, red areas are uphill and blue areas are downhill.

LEFT
**Inside Psyche**
The floor of the crater Psyche, pictured from 30 miles (50 km) away. The surface has a thin coating of soil, or regolith, and is peppered with tiny craters.

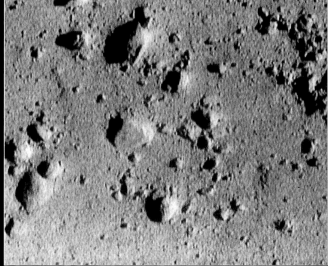

LEFT
**Landing mode**
NEAR-Shoemaker snapped this image from a range of only 800 feet (250 m) as it was descending to the surface. The large rocks shown are just a few feet high.

# 6 | The Heavenly Wanderers

**The HST keeps a watchful eye on the planets**

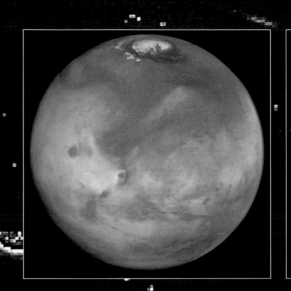

ABOVE
**Saturn spectacular**
Five pairs of images capture the movement of some of Saturn's moons in the vicinity of the ring system. Ring features visible here, from the planet outward, are the C ring, the Cassini Division, and the F ring.

INSET LEFT
**Stormy Mars**
Clouds of dust fill canyons near the equator in this image of Mars taken prior to the Pathfinder landing in summer 1997. Clouds of water-ice cover much of the north polar region.

INSET RIGHT
**Volcanic Io**
Jupiter's moon Io is one of the strangest in the solar system because of its abundance of active volcanoes. They spew out sulfur to create a colorful landscape that has been likened to the surface of a pizza.

# SCORCHED WORLDS

On many evenings of the year, we can see a bright star hanging in the darkening western sky. We call it the evening star. But star it is not—it is the planet Venus.

Of the five planets we can see in the night sky with the naked eye, Venus is the easiest to recognize. Of the others, Jupiter and Mars are easily the most distinctive, while Mercury is difficult to see because it always remains close to the Sun, and Saturn tends to merge into the stellar background most of the time. The two other planets—Uranus and Neptune—are too faint to be seen by the naked eye.

Our knowledge of the planets has been painstakingly gathered over centuries by naked-eye and telescopic observations and more recently by spacecraft, particularly those that have probed solar system space to explore planets and their moons, comets, and asteroids at close quarters.

Whereas visits by space probes are fleeting, the HST can keep a constant eye on our planetary neighbors and is always on hand to record transient and unexpected phenomena.

Ironically, observing nearby planets with the HST is a lot trickier than observing the far-distant stars. This is because the planets are moving targets, forever changing their position against the background of "fixed stars" as they orbit the Sun.

As you might expect, the two rocky planets that orbit the Sun at a closer range than Earth—Mercury and Venus—are much warmer. Temperatures on Mercury peak at over 800 degrees Fahrenheit (430°C), and those on Venus can rise as high as 900 degrees Fahrenheit (480°C).

But Mercury also experiences very low temperatures, which are due in part to the planet's slow spin. Mercury takes almost 59 days to spin around once. This slow spin, coupled with the planet's 88-day movement around the Sun, means that a point on Mercury's surface stays exposed to 176 days of sunshine at a time. This is what makes temperatures soar. On the other hand, when the Sun sets at a point on Mercury, a night begins that lasts for 176 days. Then, temperatures plummet as low as –290 degrees Fahrenheit (–180°C).

Mercury is too small to have enough gravity to hold onto an atmosphere. This lack of atmosphere not only allows the temperature extremes between day and night, but also makes the planet vulnerable to meteorites. In the distant past, lumps of rocky debris rained down on the planet and gouged out craters large and small. Today the craters remain more or less unchanged, for little erosion takes place on this airless world.

Ground-based telescopes show few details of Mercury's surface. Most of our knowledge about the planet has come from Mariner 10, a probe that made three passes of the planet beginning in March 1974. The HST does not attempt to target Mercury because it is always too close to the Sun.

RIGHT
**Pockmarked Mercury**
The images that made this high resolution mosaic were taken in January 2008, by the Messenger spacecraft as it prepared to go into orbit around Mercury. The planet is almost completely covered in craters, many being a few billion years old. They date back to the era of intense asteroid bombardment that occurred throughout the inner solar system when it was young.

## ESSENTIAL MERCURY

| | |
|---|---|
| Diameter at equator: | 3,032 miles (4,880 km) |
| Average distance from Sun: | 36 million miles (58,000,000 km) |
| Spins on axis in: | 58 days, 15 hours |
| Circles Sun in: | 88 days |
| Moons: | 0 |

BELOW
**Cratered Vista**
Close-up Mercury looks very similar to the Moon's highland regions. This strip shows a 600-mile (966-km) equatorial band stretching between the craters Boethius and Rūdaki. Eleven color filters have been used, and the color shown is greatly exaggerated to emphasize the differences in the minerals of the surface rocks and lava.

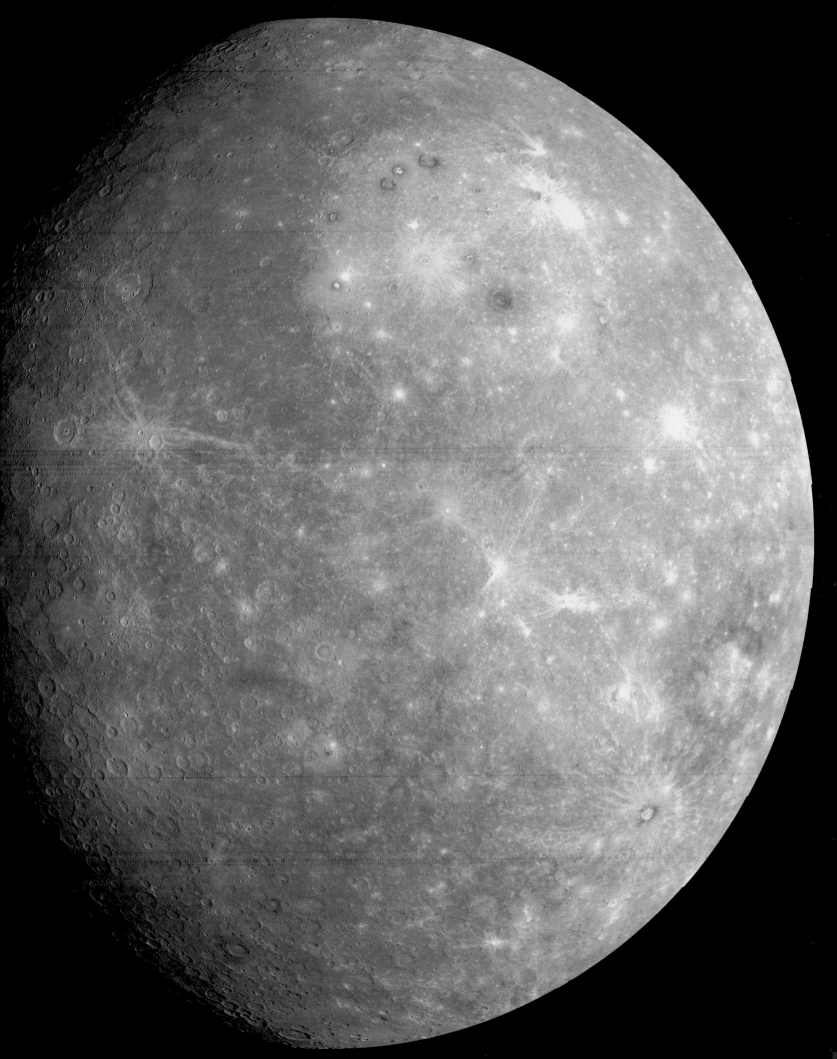

# THE PLANET FROM HELL

The planet we know as the evening star, Venus, is rocky like Mercury, but could hardly be more different in most other respects. It is much larger than Mercury, and only a few hundred miles smaller across than our own planet. And like Earth, Venus is surrounded by an atmosphere.

Venus's atmosphere is much thicker than Earth's, with a crushing pressure that is 90 times what we experience. The Venusian atmosphere is made up almost entirely of carbon dioxide, the gas that on Earth traps solar energy and causes global warming. On Venus, there is so much carbon dioxide in the atmosphere that it creates a runaway greenhouse effect that has heated up the whole planet. The temperature on Venus is over twice the temperature of your average domestic oven, a temperature that would melt metals such as tin and lead.

The thick clouds that fill the atmosphere permanently hide the planet's surface from our view. However, they are not made up of water droplets like Earth's clouds, but are made up of droplets of sulfuric acid. What a hell of a planet Venus is. If you were to go there, you would be simultaneously oven-baked, crushed, and etched to death!

## SHAPED BY VOLCANOES

What does Venus look like beneath the clouds? The Venera 9 space probe sent back the first surface picture when it landed

**LEFT**
**Veiled Venus**
In June 2007, on its way to Mercury, the Messenger spacecraft took this visual image of the planet Venus. We are looking down on to the cloud deck that completely covers the planet. The clouds consist of sulphuric acid droplets. They reflect about 75 percent of the sunlight that falls on them, making the planet very bright in the sky, but its surface murky.

**BELOW**
**Soaring volcanoes**
Maat Mons is one of the most impressive of Venus's many volcanoes. Around 4 miles (6 km) high, it is named for an ancient Egyptian love goddess. Most features on the planet have been given female names. For example, there is a plateau named Guinevere, a chasm named Diana, and a crater named Cleopatra.

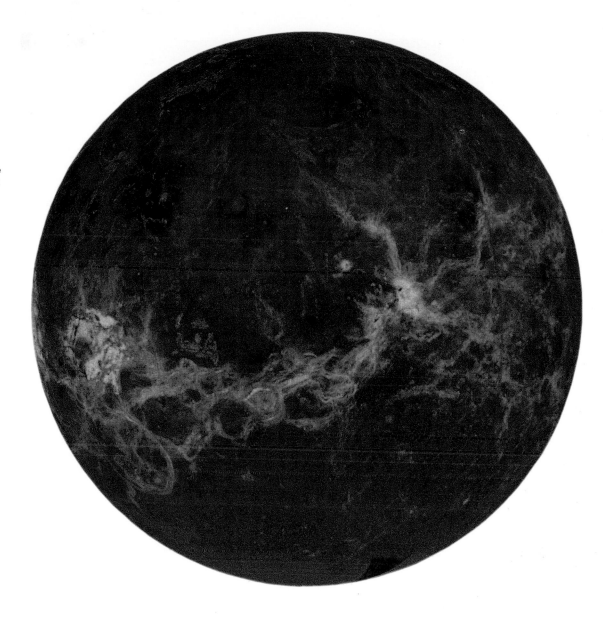

RIGHT
**Lifting the veil**
The Magellan probe used radar to penetrate Venus's cloud cover and image the surface. The bright region in this global mosaic of images is the larger of the planet's 2 continents, Aphrodite Terra.

in 1975, showing a landscape strewn with scattered rocks. Radio telescopes operating in radar mode began to reveal the general topography of the planet at much the same time. Radar can reveal surface features because it uses radio waves, which can penetrate clouds.

It was a radar probe named Magellan that completed the first high-resolution survey of Venus's surface between 1990 and 1994. Orbiting around the planet, it returned amazing images, revealing virtually a whole planet shaped by volcanic activity. Hundreds of volcanoes cover the landscape, many of them miles high, and some of them could be active.

The low-lying plains that make up more than four-fifths of the surface of Venus have been formed by repeated flows of lava spewed out from the volcanoes. Most of the volcanoes are of the shield type found widely on Earth, which have a particularly runny lava that can flow a long way.

There are only two main highland regions on Venus, which we can liken to continents. The largest, Aphrodite Terra, lies near the equator and is about as big as Africa. The other, Ishtar Terra, is about Australia's size and lies to the north.

Venus, like Mercury, always stays relatively close to the Sun. The HST has only made occasional observations of the planet, when Venus has been at its greatest elongation (its farthest point to the east or west of the Sun).

The HST hasn't observed the surface of Venus because it doesn't operate at microwave or radar wavelengths. It has, however, revealed features of the atmosphere by observing at ultraviolet wavelengths.

## ESSENTIAL VENUS

| | |
|---|---|
| Diameter at equator: | 7,521 miles (12,104 km) |
| Average distance from Sun: | 67 million miles (108,000,000 km) |
| Spins on axis in: | 243 days |
| Circles Sun in: | 224 days, 16 hours |
| Moons: | 0 |

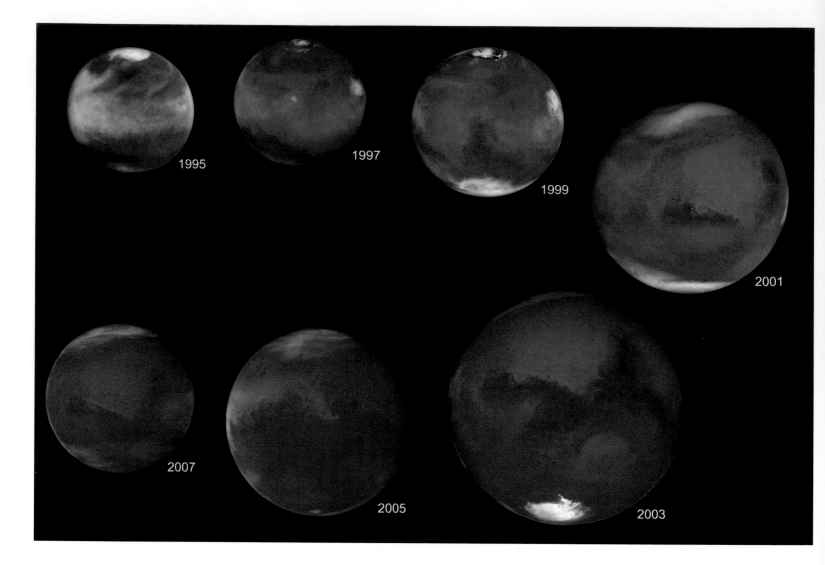

# THE RED PLANET

After Venus, two planets vie for the title of the brightest: Mars and Jupiter—the next two planets beyond Earth, going away from the Sun. They can both shine more brilliantly than any night-time star, but we can easily distinguish them. Whereas Jupiter shines a brilliant white, Mars has a distinct red–orange hue, which accounts for its popular name, the Red Planet.

It was the color of Mars that prompted its name in classical times. Likening its color to that of blood and fire, the Romans named it after their god of war.

Through a telescope you can see vague, dark features on Mars, and there are ice caps that come and go. Careful observation reveals that the planet rotates in a little over one Earth day. Space probes have revealed clouds in the Martian atmosphere.

An atmosphere, clouds, a day of similar length to our own, ice at the poles, dark markings—could they be vegetation? Surely this must be a planet like Earth? Maybe it also has intelligent life? Many people a hundred years ago believed this to be the case.

Prominent among these believers was U.S. astronomer Percival Lowell, who founded the Lowell Observatory in Arizona specifically to study Mars. He was convinced that the "canals" he thought he saw on the surface were being built by a dying race of Martians, trying to channel water from the ice caps to farmland near the warmer equator.

The reality is quite different. A series of space probes from 1965 to the present day—from Mariners and Vikings to Mars Express, Exploration Rovers, and Reconnaissance Orbiter—

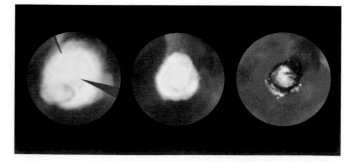

LEFT
**In retreat**
The extent of the polar ice caps on Mars changes with the seasons (which are nearly twice as long as Earth's). These HST images show how the northern ice cap shrinks over a period of five months, in between early spring and early summer.

## WATER AND WEATHER

There may be no canals on Mars, but there are channels that look much like dried-up riverbeds on Earth. They were almost certainly created by flowing water long, long ago. We know that today the only water on Mars is in the form of ice, in the polar caps and the crystals that make up the clouds. Recent spacecraft like Mars Odyssey have also detected the presence of vast quantities of water ice just below the surface—a kind of Martian permafrost.

If water did once flow on Mars, and astronomers are convinced that it did, then the planet must once have had a much milder climate than it does now—one in which water could exist as a liquid. If so, could life have gained a toehold in that milder climate? Maybe. Perhaps the first human explorers of the Red Planet will find out by unearthing fossils of ancient Martian life. We shall have to wait and see.

The HST has shed no light on the prospects of life on Mars. But it has provided a wealth of information about the Martian atmosphere and weather, closely monitoring the changes over the seasons of the Martian year, which is nearly twice as long as our own.

have shown Mars to be a barren planet, covered in vast deserts and pockmarked with craters. There is only a trace of atmosphere, and temperatures are generally much lower than those on Earth; they rise above freezing only near the equator in mid-summer. There are no signs of canals, vegetation, or life of any kind—intelligent or otherwise.

## MARS'S GRAND CANYON

Mars does, however, boast a natural linear feature—a great gash in the planet's crust that runs along the equator. This ancient geological fault system in the planet's crust is named Valles Marineris, or Mariner Valley, after the Mariner spacecraft that discovered it. It begins near the other outstanding natural feature on Mars—the bulge of the Tharsis Ridge, which carries four of the biggest volcanoes we know in the solar system.

Valles Marineris runs for nearly 3,000 miles (5,000 km), in places widening to more than 250 miles (400 km) and plunging as deep as 4 miles (6 km). This vast system of canyons dwarfs Earth's Grand Canyon in Arizona.

## ESSENTIAL MARS

| | |
|---|---|
| **Diameter at equator:** 4,220 miles (6,792 km) | |
| **Average distance from Sun:** | 142 million miles (228,000,000 km) |
| **Spins on axis in:** | 24 hours, 37 minutes |
| **Circles Sun in:** | 687 days |
| **Moons:** | 2 (Phobos and Deimos) |

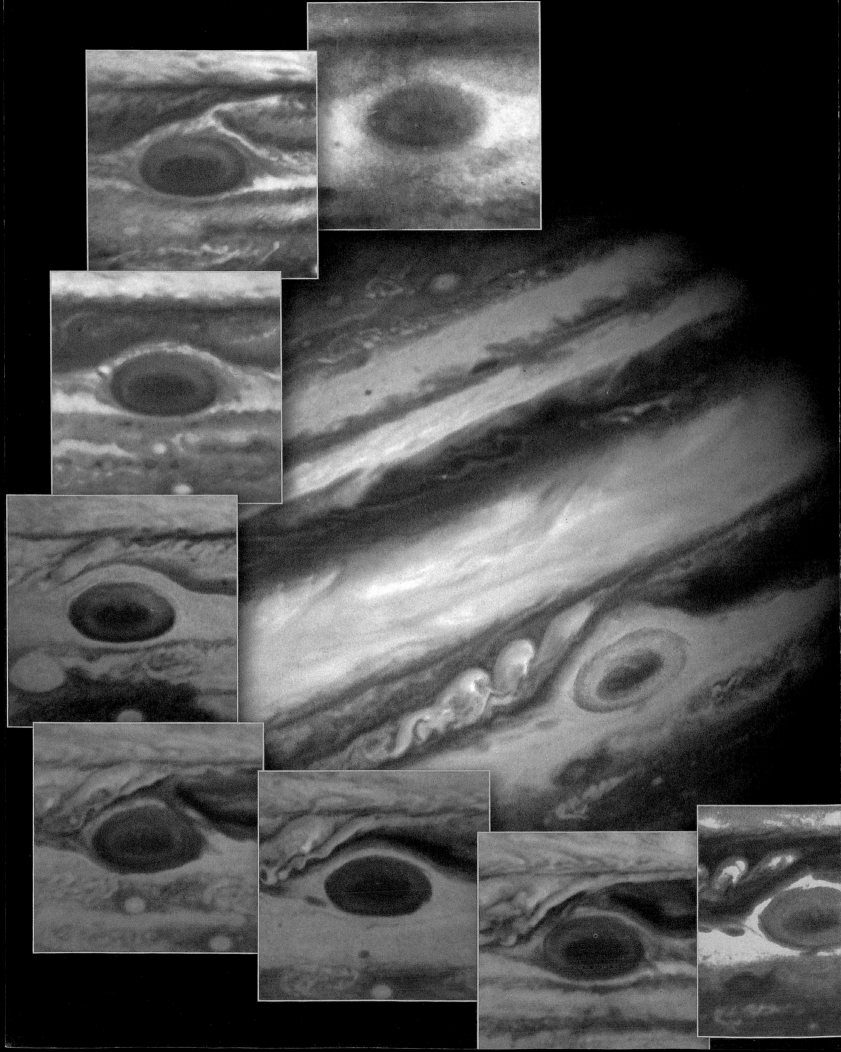

# GIGANTIC JUPITER

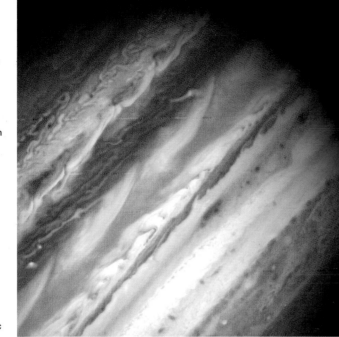

With 11 times the diameter of Earth and with more mass than all the other planets put together, Jupiter is truly gigantic. Even though it never gets closer to us than about 400 million miles (600,000,000 km), it shines brilliantly in the night sky. This is due not only to its enormous size, but also to its cloudy atmosphere, which reflects light well.

Jupiter is quite a different body from the terrestrial planets. Its atmosphere is much deeper and is composed mainly of hydrogen, with some helium. There is no solid surface, no rocky landscape, beneath Jupiter's atmosphere.

Pressures at the foot of the atmosphere rise so high that they compress the hydrogen gas into liquid, causing the whole planet to be covered by an endless ocean of liquid hydrogen. More than 12,000 miles (20,000 km) below the ocean surface, pressures soar so high that they compress the very atoms of hydrogen. The hydrogen turns into a kind of liquid metal, rather like the liquid metal mercury we find on Earth.

Saturn has a similar makeup to Jupiter, and both planets are often referred to as gas giants. So are the next two planets out, Uranus and Neptune, though they have a somewhat different composition. All four planets have only a relatively small core of rock at the center.

## THE STORMY ATMOSPHERE

Jupiter has the most fascinating atmosphere of all the eight planets. It is criss-crossed with bands of colorful clouds, and many other kinds of features—swirls, eddies, plumes, and ovals. Astronomers call the dark bands belts and the light ones zones.

The clouds have been drawn out into these parallel bands by Jupiter's rapid rotation. Strangely, the biggest planet rotates the fastest, spinning round once on its axis in a little under 10 hours. The various bands don't all travel at the same speed; those in the equatorial region travel fastest. And the winds in adjacent bands travel in opposite directions. The resulting interaction between them sets up violent eddies and turbulence that creates the wavelike features and ovals that cover Jupiter's disk.

Mostly, these features are constantly changing and transient. But one persists: the famous Great Red Spot (GRS). First glimpsed more than three centuries ago and viewed periodically ever since, the GRS is an enormous super-storm.

## RED SPOTS

Jupiter's Great Red Spot, which is about twice the size of Earth, is the largest known storm in the solar system. It is a high-pressure region that rotates counter clockwise,

| ESSENTIAL JUPITER | |
| --- | --- |
| Diameter at equator: | 88,850 miles (143,000 km) |
| Average distance from Sun: | 483 million miles (778,000,000 km) |
| Spins on axis in: | 9 hours, 55 minutes |
| Circles Sun in: | 11 years, 10 months |
| Moons: | 63 |

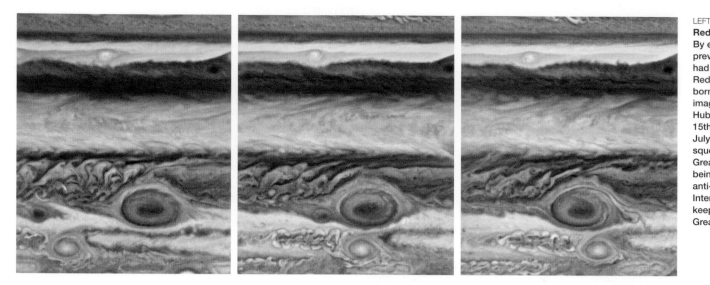

LEFT
**Red Spot and Junior**
By early 2006 a previously white storm had turned red and Red Spot Junior was born. These three images taken by the Hubble WFTC2 on May 15th, June 28th, and July 8th 2008 show it squeezing past the Great Red Spot and being caught up in its anti-cyclonic spin. Interactions like these keep powering the Great Red Spot.

completing one rotation every six days. Over longer periods of time its shape, size, and color can be seen to change.

A more recently formed and smaller storm became known as Red Spot Junior because of its similarity to the Great Red Spot. Originally three white oval storms that had merged, it turned red over late-2005 to early-2006. Like its bigger counterpart it was also in the planet's southern hemisphere.

Hubble observed the two spots as they passed each other in July 2006 and has monitored Red Spot Junior since. In Hubble images of 2008 the two were once again seen close to each other. On this occasion a third red spot, called the Little Red Spot, or sometimes Baby Spot, was also in the neighborhood. But it was short-lived; it collided with and was destroyed by the Great Red Spot in July 2008.

## JUPITER'S AURORAS

The HST has studied Jupiter's atmosphere routinely; it is capable of returning high-resolution images every hour and a half. It has studied the planet in both the visible spectrum and at invisible wavelengths. Ultraviolet studies reveal, for example, auroras—light displays that are similar to the Northern and Southern Lights seen in Earth's polar regions.

Hubble has observed Jupiter's auroras since 1990. But in late 2000 and early 2001, much was learnt about them by combining contemporaneous observations made by three spacecraft at different viewpoints. The craft were the Hubble telescope in Earth orbit, the Galileo craft in orbit around Jupiter, and Cassini on its way to Saturn. The auroras occur in the upper atmosphere of Jupiter; around the planet's magnetic north pole, and its southern counterpart. They take the form of an oval ring which is usually hundreds of miles/kilometres wide.

## CREATING AN AURORA

Jupiter has an intense magnetic field, 20,000 times greater than Earth's. This produces a magnetotail that points away from the Sun. Electrons from the solar wind can be captured in this tail region at a distance of 15 to 30 jovian radii away from

LEFT
**Jupiter's auroras**
Superimposed on a visual light image of Jupiter are two ultraviolet images of the north and south auroras taken by Hubble's Imaging Spectrograph (SIST) instrument. Jovian auroras, like Earth's are produced by charged particles spiralling in down polar magnetic field lines. These particles lose their energy in the upper atmosphere making it glow.

the planet. They can then accelerate back along magnetic field lines to smash energetically into Jupiter's upper atmosphere. This takes place about 120–250 km above the visible surface of the upper cloud layers. Here they excite molecules of ammonia, methane, hydrogen, and hydrocarbons to produce a whole range of wavelengths from the infra-red, through the visual, to the X-ray—an aurora.

Along with the auroras on Earth and Saturn, the number and brightness of Jupiter's auroras depends on the number of

BELOW
**The big four**
These Hubble images show Jupiter's four largest moons to scale. Ganymede, the largest, is even larger than Mercury; Io is volcanic and the HST has been used to monitor this activity. Hubble has found a thin oxygen atmosphere around Europa and also revealed fresh ice on Callisto.

**Ganymede**  **Callisto**  **Io**  **Europa**

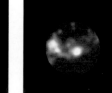

LEFT
**Ganymede**
This April 9th 2007 image of the southern hemisphere of Jupiter shows Ganymede, the largest moon in the solar system, just about to disappear behind the planet. Ganymede orbits Jupiter every seven days. The image is so clear that the white impact crater Tros is visible on the moon.

electrons in the solar wind. This varies drastically with the solar cycle of magnetic activity. Volcanic activity on Jupiter's moon Io also produces electrons that contribute to the production of Jupiter's auroras. Jovian auroras are typically 1,000 times more energetic than Earth's and they produce high-frequency radio waves that can be detected from Earth.

Interestingly, brighter regions of auroral activity are produced on the specific field lines that connect Jupiter to its inner large moons Io, Europa, and Ganymede. These all orbit well inside Jupiter's magnetosphere and can "short out" its electrostatic fields, leading to electron accelerations, which go on to produce the auroral activity.

## COLLIDING WITH JUPITER

The HST has the capability to look at Jupiter on demand when something unexpected happens. In late July 2009 the telescope was undergoing a checkout and calibration of its Wide Field Camera 3 that had been installed only recently on the telescope's final service mission. The work was put on hold when an expanding mark appeared on Jupiter.

The mark had formed when a small object, probably an asteroid, collided with the planet; plunging into Jupiter's atmosphere and disintegrating. This event occurred during the same week but 15 years later than the collision of Comet Shoemaker-Levy 9 with the planet (see page 138), which was also witnessed by Hubble.

## GALILEAN MOONS

Jupiter has a large family of 63 moons. Most are small and at great distance from the planet. But four—Ganymede, Callisto, Io, and Europa—known collectively as the Galileans, are large worlds in their own right. All four have come under Hubble's scrutiny but Io has been picked out for special attention.

Sometimes called the pizza moon because of its colorful appearance, Io is the most dynamic moon in the solar system. Its surface is undergoing constant change due to erupting volcanoes. These volcanoes do not spew out molten rock like volcanoes on Earth, but molten sulphur instead. Hubble has been watching out for volcanic eruptions and in July 1996 caught its first one.

The HST witnessed a 250-mile-high (400-km-high) plume of gas and dust erupting from Pele, one of Io's most powerful volcanoes. Hubble's view was the first time that Pele had been seen to erupt since March 1979 when the Voyager 1 spacecraft recorded a 185-mile-high (300-km-high) eruption.

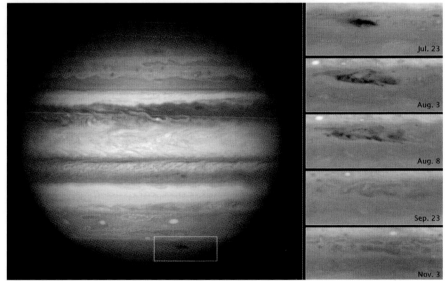

Jul. 23
Aug. 3
Aug. 8
Sep. 23
Nov. 3

ABOVE
**Asteroid Impact**
The large image shows a dark dusty cloud in the south polar region of the jovian atmosphere, on July 23rd 2009. It is thought that an asteroid about 1,600 feet (488 m) across burned up and broke up in Jupiter's upper atmosphere. The five close-ups were taken between July 23rd and November 3rd and show how the cloud speedily dispersed.

# THE RINGED WONDER

Encircled by a set of bright shining rings, the gas giant Saturn remains a firm favorite among astronomers. The other gas giants (Jupiter, Uranus, and Neptune) also have ring systems, but they are but a pale imitation of Saturn's and cannot readily be seen from Earth.

   With a similar makeup to that of Jupiter, Saturn has a banded atmosphere, but the parallel cloud bands are much less distinctive. Not so much activity takes place in the atmosphere, though furious storms do rage from time to time.

## RING-A-RING
Saturn's rings reflect light brilliantly and make the whole planet much brighter than it otherwise would be. In the night sky, it can outshine all the stars except Sirius and Canopus, but for only

some of the time. Even when it is farthest away from us—at a distance of nearly 900 million miles (1,400,000,000 km)—it still competes with bright stars such as Spica and Aldebaran.

   The amount of extra brilliance provided by the rings varies. Because of the inclination of Saturn's spin axis, on Earth we see the rings at different angles during the almost 30 years it takes Saturn to orbit the Sun. Roughly every 15 years, the rings all but disappear, when they appear edge-on to our line of sight. The rings were edge-on in 2009 and will be again in 2025.

## RINGS AND SHEPHERDS
In telescopes, we can make out three main rings from Earth: the A, B, and C rings, from outer to inner. Brightest is the B ring, which is separated from the A ring by the apparently

ringless Cassini Division. These bright rings extend to about 50,000 miles (80,000 km) beyond Saturn.

Spacecraft, such as the two Voyagers and Cassini, have changed our perspective on the rings. They, and most recently the space-based Spitzer telescope, in 2009, have detected rings inside and beyond those visible from Earth. Each set of rings is made up of thousands of individual ringlets. The ringlets are our perception of particles of material whizzing in orbit around the planet at high speed. These particles are made up mainly of ice, which is why they reflect light so well. Some particles are as small as sand grains; others are as big as boulders.

The camera eyes of the Voyager probes provided an explanation of why the particles remain in the rings. They detected a number of tiny moons among the rings that by their gravity seem to keep the ring particles in place. They are called shepherd moons, because they seem to "herd" the particles like a shepherd herds sheep.

As with Jupiter, HST observers have concentrated on monitoring the changes in Saturn's atmosphere. They have logged the progress of many a storm, after being alerted by astronomers on the ground. This is an example of the growing cooperation between terrestrial and space astronomers.

The HST has also set its sights on Saturn's planet-sized moon Titan. Bigger than Mercury, Titan is unique among moons because of its thick atmosphere, which astronomers believe might resemble Earth's early atmosphere in its chemical composition.

ABOVE
**Ring particles**
Color is used in this Cassini image of Saturn's rings to represent ring particle size. Purple indicates particles over about 2 inches (5 cm). Blue indicates particles smaller than ⁶⁄₁₀ inch (1 cm). Green indicates those in between.

LEFT
**Titan's surface**
A world of rock and ice, Titan's surface was shaped by Earth-like processes. The central feature in this Cassini image of December 2006 may be an old impact basin. Mountain ranges are to its south east.

RIGHT
**Titan's atmosphere**
Haze in Titan's atmosphere (blue), which is mainly composed of the gas nitrogen, like Earth's. Surprisingly, the pressure of the atmosphere is 50 percent higher than it is on Earth.

## ESSENTIAL SATURN

| | |
|---|---|
| Diameter at equator: | 74,900 miles (120,500 km) |
| Average distance from Sun: | 888 million miles (1,429,000,000 km) |
| Spins on axis in: | 10 hours, 39 minutes |
| Circles Sun in: | 29 years, 6 months |
| Moons: | 62 |

RIGHT
**Saturn transits**
This 2009 Hubble
image of Saturn,
when the planet
was 775 million miles
(1,247,241,600 km)
away, shows four
transiting moons. The
huge moon Titan is at
top with its shadow
above and to the right.
Just above the rings
on the left can be
seen, from left to right,
Enceladus, Dione, and
Mimas, against the
disk of the planet.
Also just above the
rings can be seen the
ring shadow, cast on
to Saturn's clouds.

LEFT
**Infrared Saturn**
To celebrate the HST's eighth birthday, the science team "gift wrapped" everybody's favorite planet in vivid colors. It was imaged in the infrared by NICMOS—the first time this instrument had been turned on the planet.

RIGHT
**Ring-plane crossing**
Two images of Saturn taken three months apart. The upper image (August 1995) records the ring-plane crossing—the time when the plane of the ring system crosses our line of sight; the last crossing took place in 2010.

LEFT
**Saturn's auroras**
Saturn's rings are seen edge-on twice in every 29.5-year orbit it makes around the Sun. This Hubble ultraviolet image is from early 2009. It also shows auroras at both of Saturn's poles. The northern auroral oval is slightly smaller than the southern one, and more intense.

# NEW WORLDS

On the night of March 13th 1781, William Herschel was studying the stars in Gemini at his home in Bath, England. German-born, he had neglected his profession as a musician to become an enthusiastic astronomer. Among Gemini's stars he came across an object that intrigued him. "A curious nebulous star or perhaps a comet," he recorded. But this body was no star or comet; it was a new planet later named Uranus.

When astronomers determined the planet's orbit, they were astonished. This new wandering star pursued an orbit that took it 1,800 million miles (2,900,000,000 km) from the Sun. This was twice as far away as Saturn, regarded since ancient times as the farthest planet. Herschel's discovery literally doubled the size of the known solar system, and triggered the search for more new worlds.

## SMALL GIANT

Uranus is the third-biggest planet after Jupiter and Saturn, but it is less than half the size of Saturn. The strangest thing about the planet is that its spin axis is tipped right over. It is tipped by 98 degrees to the plane of Uranus's orbit around the Sun, making the planet appear to roll along its orbital path. The odd angle is the result of a collision with a large asteroid when the planet was young.

Uranus appears as a featureless, blue–green orb. Like the other gas giants, its deep atmosphere is made up mainly of hydrogen and helium. There are also significant traces of methane, one gas we use to cook with on Earth. It is this gas that is responsible for the color of the atmosphere, because it absorbs red wavelengths from sunlight, producing a blue color.

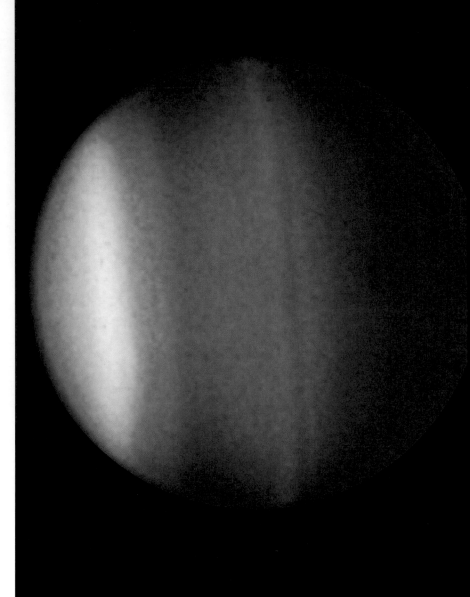

## SURPRISE, SURPRISE

In 1977, Uranus sprang another surprise. Astronomers setting out to chart the occultation of a star by the planet discovered rings around it. In 1986, Voyager 2 pictured the rings more clearly, showing that there are 11 in total. The HST discovered a second set of two giant rings in September 2004.

| ESSENTIAL URANUS | |
| --- | --- |
| Diameter at equator: | 31,760 miles (51,120 km) |
| Average distance from Sun: | 1,786 million miles (2,875,000,000 km) |
| Spins on axis in: | 17 hours, 14 minutes |
| Circles Sun in: | 84 years |
| Moons: | 27 |

Another intriguing Voyager discovery came when the probe closed in on Miranda, the smallest Uranian moon we can see from Earth. Miranda's surface displays geological features that are unique in the solar system. Quite different landscapes abut one another with no gradual transition between them.

## PATCHWORK LANDSCAPE

It is possible that such geological patchwork might have come about as a result of a catastrophic collision with another body eons ago. The collision would have smashed the moon to pieces, and then gravity would have caused the moon to reform again, with all the pieces of different geology jumbled together haphazardly.

Voyager 2 also spied a number of tiny moons, invisible to us from Earth. Among them were two shepherd moons that orbit on either side of the outer, Epsilon ring. Named Cordelia and Ophelia, they are both only about 20 miles (30 km) across.

The HST has carried out regular observations of Uranus, its rings and its moons. It has spied more activity in the atmosphere than did Voyager 2, following the changing patterns of clouds and haze. Hubble images not only uncovered the two new rings in 2004 but also two tiny moons Mab and Cupid a year earlier.

## BLUE PLANET

The enthusiastic search for other planets triggered by Herschel's discovery of Uranus came to fruition in 1846. On September 23 of that year, German astronomer Johann Galle discovered planet number eight, which was subsequently called Neptune. It proved to be only a fraction smaller than Uranus. Recent observations have shown that the two planets are remarkably similar, even though their orbits are more than a billion miles (1,600,000,000 km) apart.

Neptune has relatively more methane in its hydrogen/helium atmosphere, and as a result the planet appears a deeper blue than Uranus. But both planets seem to have a similar ocean of warm water, liquid ammonia, and methane underneath the atmosphere.

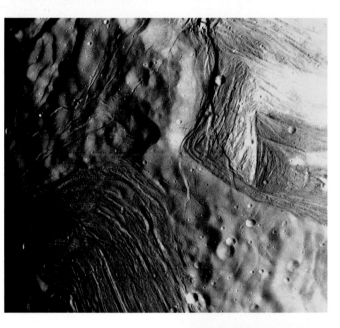

ABOVE
**Epsilon's ringlets**
Many individual ringlets make up Uranus's outermost ring, the Epsilon, pictured here by Voyager 2. The Epsilon is also the broadest ring, measuring about 60 miles (100 km) across.

ABOVE
**Amazing Miranda**
The strange patchwork landscape of Miranda, where different kinds of terrain abut—cratered hilly regions, chevron or V-shaped areas, and oval grooved regions that look like a gigantic racetrack. No geology exists like it anywhere else in the solar system.

RIGHT
**New rings**
The HST's sharp view has uncovered a pair of faint giant rings girdling Uranus and its previously known ring system. It has also spied two new moons, Mab and Cupid.

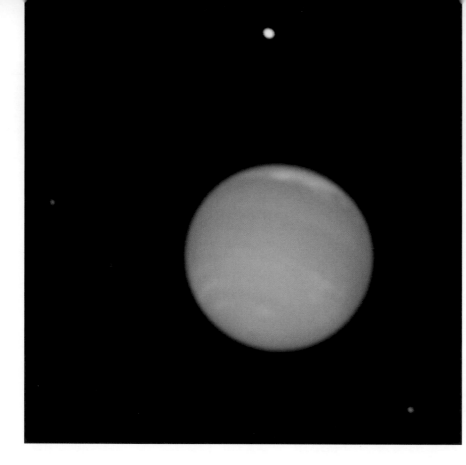

**Natural world**
This natural color view of Neptune was produced by combining HST images in red, green, and blue light. It is one in a series taken in 2005 that show a dynamic atmosphere. Clockwise from the top are four of Neptune's moons: Proteus, Larissa, Despina, and Galatea.

RIGHT
**Changing weather**
The HST monitors the rapidly changing weather on Neptune. Atmospheric features become visible in enhanced-color views (right) and in images taken using methane filters (left).

## NEPTUNE'S CLOUDS AND STORMS

When Voyager 2 flew past Neptune in 1989, it revealed that the blue planet's atmosphere was much more active—had more weather—than Uranus's. Being much farther from the Sun, Neptune receives less solar energy, which is usually what drives planetary weather systems. So you might expect that Neptune's atmosphere would have little weather.

The truth is quite the reverse. Voyager's cameras spotted distinct bands in the atmosphere, rather like Jupiter's and Saturn's. There were also oval storm-type features; a particularly large one became known for obvious reasons as the Great Dark Spot. There were also high-level, wispy clouds, rather like the cirrus clouds we get in Earth's atmosphere. But Neptune's wispy clouds seem to made up of crystals of methane rather than water ice.

Why is Neptune's atmosphere more active than that of Uranus? It seems that the planet has an internal heat source, which drives the planet's weather systems. Internal heating would also explain why at the cloud tops, both Uranus and Neptune have more or less the same temperature, of about −350°F (−210°C).

### ESSENTIAL NEPTUNE

| | |
|---|---|
| **Diameter at equator:** | 30,780 miles (49,530 km) |
| **Average distance from Sun:** | 2,800 million miles (4,505,000,000 km) |
| **Spins on axis in:** | 16 hours, 6 minutes |
| **Circles Sun in:** | 165 years |
| **Moons:** | 13 |

## NEPTUNE'S RINGS AND MOONS

Like Uranus, Neptune has a ring system, but it is even sparser and fainter. Two of the rings are relatively bright, and the other three very faint indeed. The two brightest are named Adams and Leverrier, after the two mathematicians who independently figured out where the eighth planet could be found (Englishman John Couch Adams and Frenchman Jean Urbain Leverrier). There is also a sixth partial ring.

On its flyby of Neptune, Voyager 2 discovered six new moons, adding to the two that can be seen from Earth, Nereid and Triton. One was Proteus, bigger than Nereid but impossible to see from Earth, because it orbits so close that it gets lost in the planet's glare. Of the remaining five new moons Voyager discovered, four orbit within the ring system, presumably acting as shepherds for the ring particles.

When the HST first trained its instruments on Neptune, it revealed no trace of the Great Dark Spot Voyager had seen. But it spotted plenty of other cloud activity in the atmosphere.

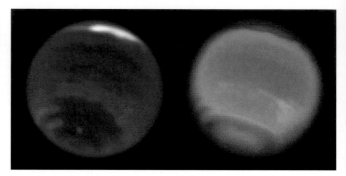

## ICE WORLDS

With Neptune, we come to the end of the progression of gas giants that dominate the outer reaches of our solar system. Then, as we penetrate ever deeper into space, we encounter planetary bodies that we call the ice worlds.

We don't even have to leave Neptune to find one. Neptune's largest moon Triton is one. Measuring some 1,700 miles (2,700 km) in diameter, it is made up of rock and ice. Its surface is covered with frozen gases, such as nitrogen and methane, as well as ice. The most curious feature of Triton's surface is its volcanoes. Totally unlike Earth's molten lava volcanoes, those on Triton emit a kind of slushy, frozen gas-ice in a process termed cryovolcanism. Geyser-like eruptions also puff out gas mixed with dark particles, which form dark streaks on the pale ice and snow.

## PLUTO

Astronomers still continued to try to find new planets long after Neptune had been discovered. Prominent among them was Percival Lowell, noted for his belief in intelligent Martian life. At the Lowell Observatory in Arizona, Clyde Tombaugh eventually found a planetary body in February 1930. It was named Pluto.

Pluto was considered the ninth planet of the solar system from its discovery until 2006. It was then the smallest and most distant planet. But during the closing decades of the twentieth century, its planetary status was increasingly questioned by some astronomers.

This small rock-and-ice body is unlike its large, gassy planetary neighbors. It follows a very elongated orbit and it is inclined to the orbital plane of the planets. And for 20 years of its 248.6-year orbit Pluto is closer to the Sun than Neptune. This last occurred between 1979 and 1999. Pluto also has a moon, Charon, that is about half its own size.

Pluto has not been visited by spacecraft and is comparatively little known. New Horizons is on its way and will arrive in 2015. Pluto is much too remote for ground-based telescopes to spy surface features. But spectroscopic analysis has shown that it is covered with frozen nitrogen, methane, and other gases. The HST has been invaluable in the study of Pluto. It can separate Pluto and Charon, and in May 2005 it discovered two additional small moons, Nix and Hydra.

## DWARF PLANETS

In October 2003, an object larger than Pluto was found, and astronomers felt that Pluto could no longer be regarded as a planet. In August 2006, a new class of object—called dwarf planets—was introduced by the International Astronomical Union, the astronomer's professional world body, and Pluto was to be the prototype.

Dwarf planets, like planets, orbit the Sun, and have sufficient mass and gravity to be nearly spherical, but, unlike a planet, a dwarf planet has not cleared the neighborhood around its orbit. There are five known dwarf planets and more

ABOVE
**Pluto and moons**
This Hubble image is one of the best of Pluto and its moons. Its largest moon Charon, discovered in 1978, is to Pluto's lower right. Farther to the right are the two smaller moons, Nix (top) and Hydra, discovered in 2005.

ABOVE RIGHT
**Eris**
The dwarf planet Eris and its satellite Dysnomia (left of Eris) can be seen in this August 2006 HST image. Eris is a little larger than Pluto but about three times farther away from the Sun. The mass of Eris is about 0.2 percent that of Earth.

ABOVE
**Ceres**
Visible and ultraviolet observations made by the HST are combined to produce this color image of Ceres. Brighter and darker regions could be asteroid impact features.

are expected. In addition to Pluto, there are Eris, Ceres, Haumea, and Makemake. Eris is the largest; its size is not certain but it is accepted to be between a few tens of miles and a few hundred miles larger than Pluto, which is 1,432 miles (2,304 km) across. Eris, Pluto, Haumea, and Makemake belong to the Kuiper Belt, a donut-shaped belt of rock and ice bodies encircling the solar system beyond the orbit of Neptune. The first of these were discovered in the 1990s; more than 1,000 are now known and it is expected there are vast numbers more.

Ceres is the largest asteroid and orbits the Sun within the Main Belt between Mars and Jupiter. It is made mainly of rock with water ice. The Dawn spaceprobe is scheduled to arrive at Ceres in 2015 and reveal this world to us.

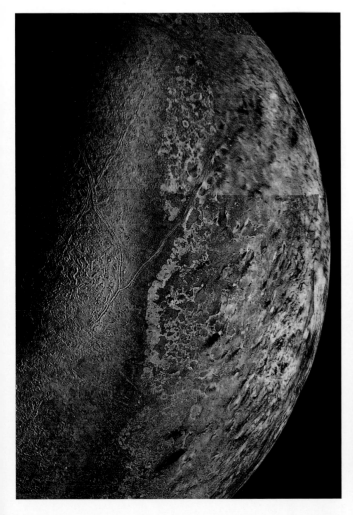

FAR LEFT
**Icy geysers**
Neptune's biggest moon Triton is in deep freeze. It has the coldest surface (–390°F/ –235°C) of any body we know in the solar system. The image shows the south polar region, which seems to be covered with pinkish snow. The dark streaks show where fine dust erupts from icy geysers.

LEFT
**Pluto revealed**
The dwarf planet Pluto is a tough target even for the HST. The resolution of a few hundred miles shows a surface of white and dark yellow patches on a charcoal black background. Seasonal ice is forming in the cooler polar regions and in other spots methane is breaking up leaving behind a dark carbon-rich residue.

# Background Briefing

## Introducing Telescopes | The Hubble Space Telescope

ABOVE
**VLT telescopes**
The four 27-foot (8.2-m) reflectors that make up the Very
Large Telescope (VLT) at the Paranal Observatory at
Atacama in Chile. The telescopes are named for sky objects
in the native Chilean dialect. In the foreground is Yepun;
behind it (from the left) are Antu, Kueyen, and Melipal.

# INTRODUCING TELESCOPES

We can explore the Universe with just our eyes and view the constellations, meteors, comets, and eclipses. However, as an optical instrument, the eye has a great disadvantage. It has a small aperture (opening)—the pupil—to let the light through. The principal instrument that astronomers use, the telescope, has a much greater capacity for light-gathering.

The Italian genius Galileo heard of the newly invented telescope in 1609; he quickly constructed his own and trained it on the heavens. He became the first to see the mountains on the Moon and the phases of Venus. The four large moons of Jupiter that he saw are called the Galilean moons in his honor. Galileo used a set of lenses to gather and focus light from the heavens. Many amateurs still use this kind of telescope, called a refractor. An objective lens at the front gathers and focuses incoming light, and the observer views the image formed through an eyepiece.

However, most astronomers use a telescope with mirrors, called a reflector. A curved, parabolic mirror (the primary) is used to gather and focus the light into an image, which is then viewed through the eyepiece. The instrument favored by amateur astronomers is called a Newtonian reflector, because it uses the same configuration as the reflector built by Isaac Newton around 1672. A secondary mirror reflects light gathered by the primary mirror into the eyepiece at the side. In an alternative configuration called the Cassegrain, a secondary mirror reflects light into the eyepiece through a hole in the primary. This is the configuration used by the HST.

The progress of astronomy since Galileo's time is a result of the improvement in the size and design of telescopes. The bigger the telescope, the more efficient it is in gathering light, and the more clearly it can resolve (distinguish) objects.

William Herschel built the finest instruments of his day and discovered a new planet, Uranus, in 1781. Lord Rosse built a gigantic instrument he called the Leviathan and was first to spot the spiral nature of M51 (Whirlpool Galaxy) in 1845. With the groundbreaking 100-inch (2.5-m) Hooker telescope, Edwin Hubble proved in the 1920s that there were star systems—galaxies—beyond our own and that the Universe appeared to be expanding.

## VERY LARGE TELESCOPES

Today's foremost instruments use very large mirrors. The two Keck telescopes at the Mauna Kea Observatory in Hawaii have mirrors 33 feet (10 m) across. The four instruments that make up the Very Large Telescope (VLT) in Chile each have mirrors 27 feet (8.2 m) across.

The mirrors in these telescopes are not made of a single piece of glass. It would be difficult to control and adjust a mirror of that size without creating distortions. These mammoth mirrors are made in segments, which merge together to make one big mirror. Each segment is individually supported and controlled by a computer that constantly adjusts it so that, all together, the segments always form the perfect shape. This technique is known as active optics.

Individually, each of the four telescopes of the VLT is superb, and when they work together as they are designed to, they produce spectacular images that match the quality of those from the HST. When they work in unison, along with three smaller 6-foot (1.8-m) instruments, they create an effective telescope 390 feet (120 m) across.

LEFT
**Keck domes**
The domes of the twin Keck telescopes on the summit of Mauna Kea, Hawaii. Inside are two of the world's largest optical and infrared telescopes. Each is eight storeys tall and has a mirror more than 30 feet (10 m) in diameter.

**Early birth**
The Lynx Arc, a star-forming region is revealed by combining images from the Keck telescopes and from two space telescopes, the HST and Rosat. It is 12 billion light-years away and is a rare glimpse of starbirth in the early Universe.

LEFT
**The Crab Nebula**
A supernova was seen in 1054, in the constellation of Taurus. The Crab Nebula is the expanding remnant of that exploding star. This view was taken by the European Southern Observatory's Very Large Telescope (VLT), in November 1999, around 950 years after the explosion was first seen. Three images through three different filters were combined to make the picture.

# THE INVISIBLE UNIVERSE

In 1931, a communications engineer named Karl Jansky from Bell Laboratories in New Jersey, was trying to identify interference he was getting on his radio equipment. After ruling out any local sources, he realized what was happening. The interference was coming from the heavens. That observation proved to be the springboard to one of the most exciting branches of astronomy—radio astronomy. It led astronomers to discover whole new kinds of heavenly bodies, including quasars, pulsars, and radio galaxies.

Astronomers tune into the radio waves that emanate from the heavens with radio telescopes that look nothing like their optical counterparts. A radio telescope usually takes the form of a huge metal dish. The dish collects incoming radio waves and focuses them on a central antenna. The signals then pass to a receiver. After amplification, they are fed to a computer for analysis and display as a false-color image, with different colors being assigned to different radio intensities and wavelengths.

## BIG AND VERSATILE

The biggest single radio telescope is located near Arecibo in Puerto Rico. Measuring 1,000 feet (300 m) across, it is built into a natural bowl in a hilltop. Since it is fixed, it must rely on the Earth's rotation to direct it toward different parts of the sky.

A much more versatile instrument is the Very Large Array near Socorro, New Mexico. It is made up of 27 separate dishes, each 82 feet (25 m) across. The dishes are mounted on rail tracks and can be moved into various configurations. Together they present a collecting area some 20 miles (30 km) across.

Radio telescopes at different locations can also be linked to simulate even larger dishes. This is called very long baseline interferometry (VLBI). Theoretically, it can create an effective collecting area with the diameter of the Earth.

## INVISIBLE ASTRONOMY

The radio waves that pour down on Earth from the heavens come from what is often called the invisible Universe—the Universe we can't see because our eyes can't detect it. As witnessed by radio images, the invisible Universe can look nothing like the visible one. Still, there is more to the invisible Universe than just radio waves.

We detect objects in the heavens by the energy they emit. Visible light and invisible radio waves are just two forms of energy we receive from heavenly bodies such as stars and galaxies. They also give off energy as invisible gamma rays, X-rays, microwaves, ultraviolet, and infrared rays. All these rays are kinds of electromagnetic radiation, which differ only in wavelength.

ABOVE
**Parkes telescope**
One of the largest radio telescopes in the southen hemisphere, at the Australian National Radio Observatory, near Parkes, New South Wales. Its dish measures 210 feet (64 m) in diameter.

RIGHT
**Cassiopeia A**
Data from the HST is combined with observations from two more space telescopes, Spitzer and Chandra to produce a stunning image of Cassiopeia A. It is the remnant of a supernova.

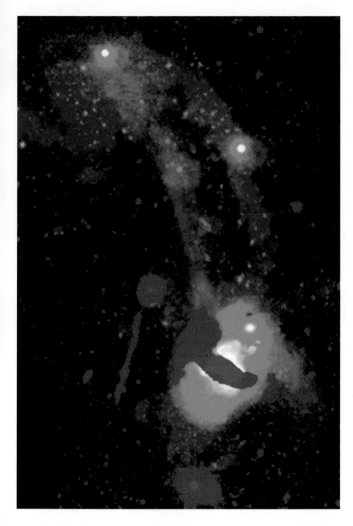

In order to get a comprehensive picture of the heavenly bodies, astronomers ideally need to study them at all of the wavelengths they give out. On the ground, astronomers can study light and radio waves from the heavens because they pass through the atmosphere. But they can't study other wavelengths because the atmosphere absorbs them. This is one of the reasons that astronomers launch telescopes and other instruments into space.

Another reason for launching telescopes into space is that, orbiting hundreds of miles high, they are above Earth's polluted atmosphere. The atmosphere, in effect, offers us a dirty window on space. It is full of dust, clouds, water vapor, and shimmering air currents, all of which distort the faint light that reaches us from the heavens.

## SPACE OBSERVATORIES

Spacecraft have been returning information about the environment from the very beginning of the Space Age. Launched on January 31st 1958, the first U.S. satellite, Explorer 1, discovered intense donut-shaped bands of radiation around the Earth, which are called the Van Allen belts. The first dedicated astronomy satellite, OAO-2 (Orbiting Astronomical Observatory-2) went into orbit 10 years later.

Since that time, scientists have launched dozens of astronomy satellites—space observatories—that span the full range of invisible wavelengths (except for radio waves). From shortest to longest, they cover gamma rays, X-rays, ultraviolet rays, infrared rays, and microwaves. The HST, while primarily used for visible light, also covers some ultraviolet and near-infrared wavelengths.

LEFT
**Radio view**
The Very Large Array reveals the peculiar galaxy Arp 299. It is the result of two spiral galaxies colliding and merging. A long tail (in blue) extends from the main body of the two.

BELOW
**Solar Dynamics Observatory**
This NASA spacecraft was launched in February 2010 onto a geostationary orbit. It will continuously observe small regions of the photosphere at multiple wavelengths. It aims to investigate the way in which the Sun's magnetic field is generated and how this field helps to accelerate solar wind particles into space.

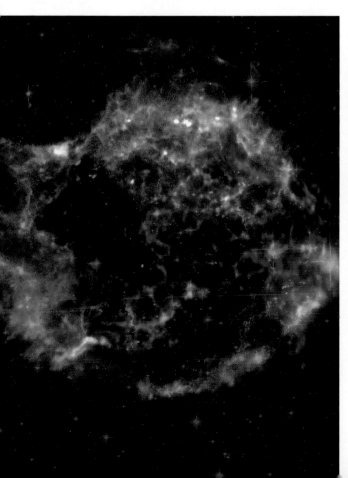

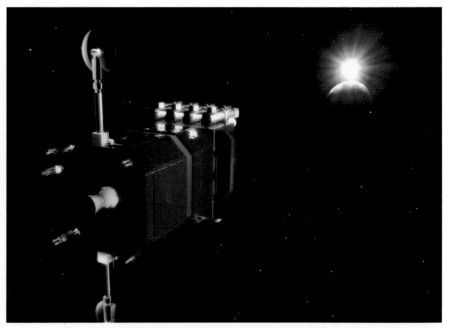

## GAMMA RAYS

Gamma rays have the shortest wavelengths and pack the most energy. They are given off by the most violent activities that take place in the Universe. Scientists have detected them coming from pulsars and quasars and in incredibly energetic bursts from unknown events in distant galaxies. Other gamma-ray sources have been identified as events in which ordinary matter and its mirror-image opposite, antimatter, annihilate each other.

The U.S. Compton Observatory, launched in 1991, provided the first comprehensive view of the gamma-ray Universe. Europe's Integral (International Gamma-Ray Astrophysics Laboratory), is designed to provide high-resolution imaging of gamma-ray sources. It operates in a looping orbit that takes it as far as 95,000 miles (153,000 km) from Earth. In this orbit it spends most of the time outside the potentially damaging Van Allen radiation belts in a relatively undisturbed space environment that is ideal for long-duration, real-time observations. Integral was launched in 2002 on a two-year mission, and was extended to December 2014.

## X-RAYS

The wavelengths of X-rays are longer than those of gamma rays and shorter than those of ultraviolet rays. They still pack a lot of energy, which is why they can penetrate body tissue. In the heavens, X-rays are emitted by very hot objects—with temperatures of millions of degrees. Typical X-ray sources include the solar corona (the outer atmosphere of the Sun), supernova remnants, and the gas that spirals around black holes. Key X-ray observatories include Rosat (Roentgen satellite, launched in 1990), Chandra (1999), and XMM-Newton (1999).

## ULTRAVIOLET RAYS

Ultraviolet wavelengths are just shorter than those of the violet light of the visible spectrum. Some ultraviolet (UV) gets through Earth's atmosphere—it is the radiation that tans us. Thankfully, most UV is absorbed by a layer of ozone in the upper atmosphere. The hottest stars are best observed in the ultraviolet because they give out most energy at these wavelengths. The International Ultraviolet Explorer (launched in 1978) provided outstanding data for 18 years. GALEX (Galaxy Evolution Explorer), launched in 2003, is investigating how star formation evolved from the early Universe up to the present. It is also identifying objects for further study.

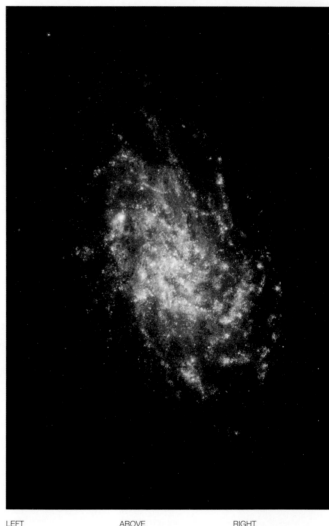

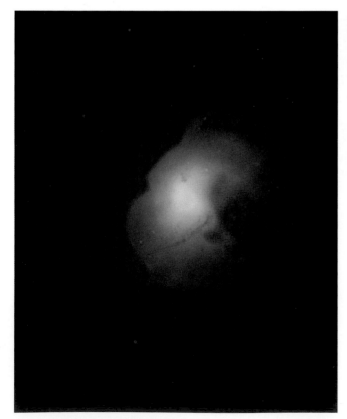

**LEFT**
**Black hole**
An X-ray image taken by Chandra of galaxy NGC 4696 shows a vast cloud of hot (red) gas surrounding a bright area that harbors a massive black hole.

**ABOVE**
**Triangulum Galaxy**
A combination of two images, one from the Spitzer Space Telescope (in the infrared), and the other from GALEX (in the ultra violet) show heated dust and young stars in the Triangulum Galaxy. It is a spiral galaxy and part of the Local Group.

**RIGHT**
**Whirlpool**
X-ray and ultraviolet images taken by XMM-Newton are combined in this view of the Whirlpool Galaxy (M51). Red indicates low energy, blue, high. Compare it with the HST view on page 94.

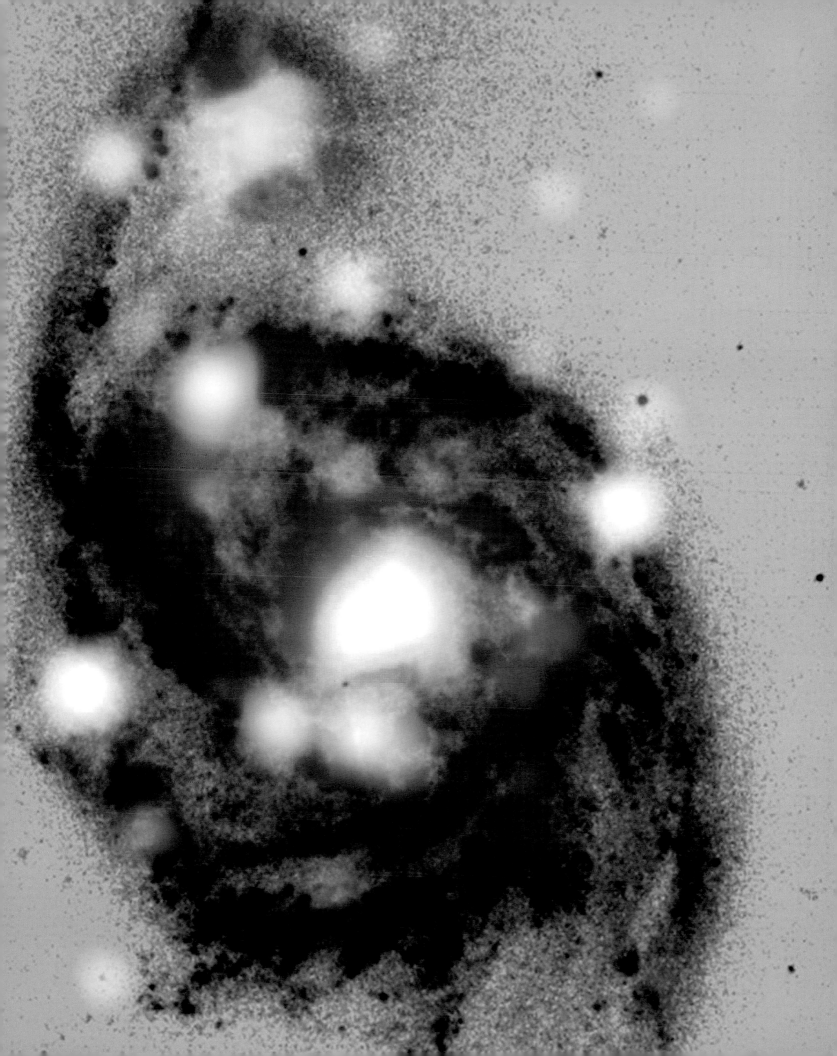

## INFRARED RAYS

Infrared rays have slightly longer wavelengths than visible red light. We receive heat from the Sun in the form of infrared rays. Some infrared wavelengths reach the ground and can be studied from mountaintop observatories. But most infrared wavelengths are completely absorbed by the atmosphere. Airborne observatories such as SOFIA (Stratospheric Observatory for Infrared Astronomy) study the infrared wavelengths from high altitudes. But the most spectacular work has been done by IRAS (Infrared Astronomy Satellite, launched in 1983), ISO (Infrared Space Observatory, launched in 1995) and Spitzer, which, since its launch in August 2003, has been studying a wide variety of astronomical objects, from the solar system to the distant reaches of the early Universe.

## MICROWAVES

Microwaves have the shortest radio wavelengths. They are used terrestrially to relay communications signals and in cooking with microwave ovens. COBE (Cosmic Background Explorer), launched in 1989, and then its successor WMAP (Wilkinson Microwave Anistropy Probe), launched in 2001, used microwaves to survey the cosmic background radiation of the whole sky. COBE detected slight variations in the radiation, regarded as the thermal "echo" of the Big Bang. WMAP measured these temperature differences and made other observations to better understand the early history of the universe.

## SOLAR OBSERVATORIES

By studying the Sun, we find out what ordinary stars are like—there are billions of stars like it in our Galaxy alone. There are also more pressing reasons for studying this, our local star, because what happens in the Sun can affect Earth and change our "space weather."

We study the Sun both from the ground and from space. On the ground, astronomers use solar telescopes that use mirrors to project an image of the sun. The McMath solar telescope at Kitt Peak Observatory in Arizona is the world's biggest, with a distinctive triangular design.

The astronauts on the Skylab space station made the first comprehensive study of the Sun from space in 1973–74. Using a series of eight instruments, they probed the Sun at many different wavelengths and achieved spectacular results. More recently, Solar Max (short for Solar Maximum Mission) was launched in 1980 to monitor the Sun at a time of solar maximum, when sunspots and other solar activities reach their climax.

Ulysses (launched in 1990) embarked on a circuitous route via Jupiter to spy on the polar regions of the Sun, which had never before been investigated. SOHO (Solar and Heliospheric Observatory) began continuous observations of our star in 1995. The Sun is also being observed by Hinode, and two craft together known as STEREO, both since late 2006, and by the Solar Dynamics Observatory since early 2010.

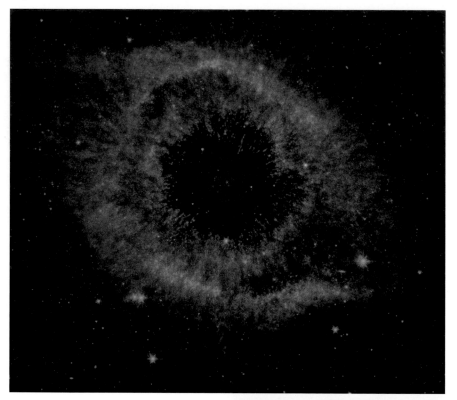

ABOVE
**Helix Nebula**
In Spitzer's infrared view of the Helix Nebula, the red center denotes the final layers of gas blown off by the central dying star. Infrared light from the outer layers of gas is represented in blues and greens.

RIGHT
**Big Bang temperature**
A whole sky map produced from seven years data taken by WMAP. The color range represents a change in the microwave background temperature of only 400 micro degrees Celsius. These fluctuations were the seeds that eventually grew into galaxies.

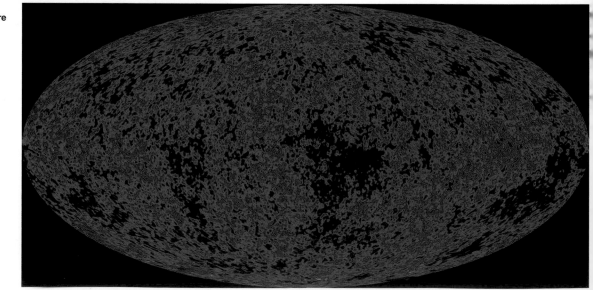

RIGHT
**Solar reflections**
The 11-story-high McMath solar telescope at Kitt Peak National Observatory in Arizona. Sunlight captured by a mirror (heliostat) on top is reflected down the inclined tunnel deep underground, where it forms an image in an observation room.

LEFT
**Hinode's Sun**
Hot gas leaps high above the Sun in this view taken by the Solar Optical Telescope aboard the Hinode spacecraft, in November 2006. The gas rises from a sunspot on the Sun's surface.

TELESCOPES WORKING TOGETHER

**LEFT**
**Radio Deep Field**
The yellow contours on this Hubble Deep Field image represent radio wave emissions observed by all twenty seven 82-foot (25-m) diameter radio dishes of the Very Large Array telescope in New Mexico, U.S.A.

Huge advances have been made in astronomy in the last sixty years by combining visual data with observations made at other, longer, and shorter wavelengths. And Hubble astronomy is no exception. Take the iconic Hubble Deep Field image of 1995 (see page 107) as an example. This picture of some of the most distant galaxies in the Universe only covers 1/28,000,000 of the sky, a mere 5.3 square arc minutes in the constellation of Ursa Major. The region was specifically chosen so that follow-up investigation could be conducted by the huge twin 33-foot (10-m) Keck telescopes on Mauna Kea, Hawaii; by the telescopes at the Kitt Peak National Observatory, Arizona; and by the Very Large Array radio telescope in Socorro County, New Mexico, U.S.A.

It is a similar story for the other deep field views that followed it, in 1998 and 2003–4 (see page 102). The choice of the deep field positions was also governed by the fact that bright nearby stars, and intense X-ray, ultraviolet, and infrared emissions were not present in these regions of the sky. This meant that objects in the deep field views could subsequently be studied at many other wavelengths.

## A MATTER OF TEMPERATURE

By collecting and analyzing the whole range of radiation emitted by astronomical objects we get a deeper and more complete understanding of such objects. It is all a matter of temperature. Stars with surface temperatures of around 10,000 degrees Fahrenheit (5,500°C), like the Sun, emit most of their radiation in visual wavelengths. The corona of the Sun has a temperature of around 3,600,000 degrees Fahrenheit (2,000,000°C) and here most of the energy is emitted as X-rays. For regions with temperatures between these two limits, the ultraviolet is very useful. Anyone interested in the cooler gas and dust regions that are condensing to form new stars turns to the infrared. And the huge cold clouds of hydrogen and other molecules that occupy the arms of the galaxies are best mapped by using their millimeter and radio wave emissions.

## STUDYING GALACTIC EVOLUTION

One of Hubble's major achievements has been its insight into galactic evolution. Its deep field observations reveal the range of galaxy types that were present in the early days of the Universe, soon after the Big Bang. If we look around the nearby cosmos, close to the Milky Way Galaxy, we see the range of galaxies some 13.7 billion years later. Originally we expected the distributions to be very different. But they are not.

The Spitzer Space Telescope works in the infrared and detects wavelengths that are five to fifteen times longer than those detectable by Hubble. Due to the large redshifts of these distant galaxies, this means that Spitzer is more sensitive to the redder stars. The brightnesses observed by Spitzer indicate that these galaxies are much more massive than expected. Instead of the ancient deep field galaxies being of low mass, and then growing by collisions to produce the giant galaxies we see today, it seems that there were huge giant galaxies in these early days too.

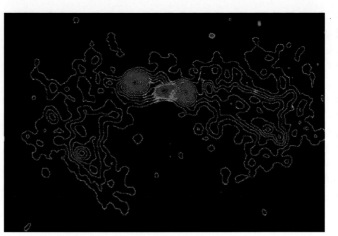

LEFT
**An active galaxy**
The radio intensity of the lobes of an active galaxy in the Hubble Deep Field is represented by yellow contour lines. These lobes are produced by beams of particles ejected by the galaxy's nucleus. The radio image was produced by combining data from both the U.K.'s MERLIN and the U.S.A.'s VLA radio telescopes.

## STAR CLUSTER

The star cluster NGC 4755, popularly known as the Jewel Box has been looked at by the HST and two European Southern Observatory (ESO) telescopes in Chile, South America. These are the Very Large Telescope on Cerro Paranal (see page 166), and the 7-foot (2.2-m) at ESO's La Silla Observatory.

The Jewel Box cluster consists of stars born only about 16 million years ago from the same cloud of gas and dust. It is easily seen and imaged from Earth but the Hubble can collect light of shorter wavelength than is possible on the ground. The HST observations, ranging in the far-ultraviolet to the near infrared, revealed details not seen before. The dimmest stars captured in the Hubble image are less than half the mass of the Sun.

BELOW
**The Jewel Box**
The open star cluster NGC 4755 is also known as the Jewel Box. It has several blue–white super-giant stars and a solitary ruby-red supergiant (above center). The brighter stars are about 20 times the mass of the Sun.

## GREAT OBSERVATORIES

In order to study and more fully understand space objects, NASA built four space telescopes collectively known as the Great Observatories. They were designed to each work in different wavelengths (visible, gamma ray, X-ray, and infrared). Importantly they were to be launched so that all four of them would work in space at the same time for at least part of their lifetimes. In this way astronomers could make simultaneous observations of an object at different wavelengths.

The first of the Great Observatories to be launched was the Hubble, in 1990. Second was the Compton Gamma Ray Observatory, launched in 1991, and then came the Chandra X-ray Observatory, in 1999. Last was the infrared telescope, the Spitzer Space Telescope, launched in 2003. Although all four should have been launched by a space shuttle orbiter, Spitzer made it into space aboard a Delta II rocket. Three of the telescopes—the HST, Chandra, and Spitzer—continue to live up to their name, regularly returning truly great observations. Unfortunately, a gyroscope failed on the Compton and the telescope was deliberately de-orbited in 2000, with the majority of it burning up in Earth's atmosphere.

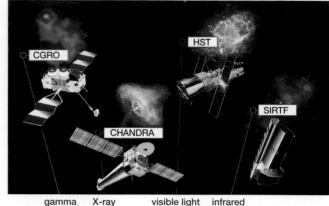

gamma    X-ray          visible light    infrared

## OBSERVING SUPERNOVA REMNANTS

The use of multiple space telescopes is extremely rewarding when looking at supernova remnants. A perfect example is Kepler's supernova remnant. Johannes Kepler, the famous German theoretical astronomer who worked with Tycho Brahe saw a supernova in the constellation of Ophiuchus in 1604.

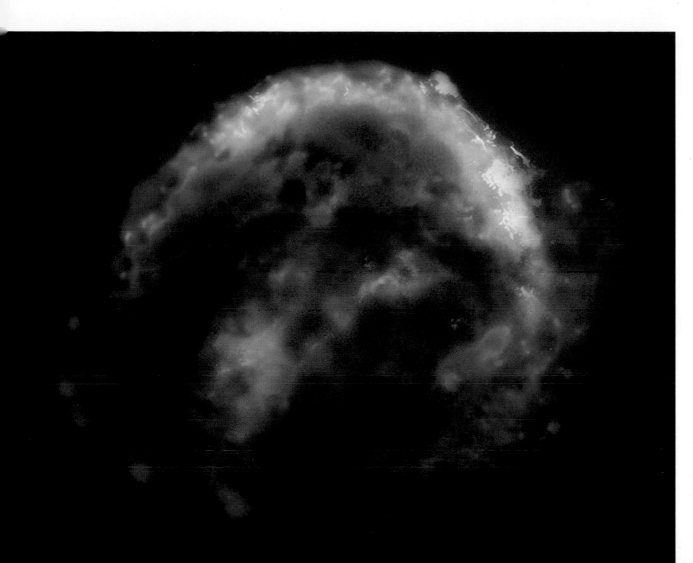

**Kepler's remnant**
The large composite image shows the expanding remnant of the supernova seen by Kepler in 1604; Chandra data is blue–green, Hubble yellow, and Spitzer red. The remnant is now 14 light-years across and is still expanding by a few million miles/ kilometers per hour.

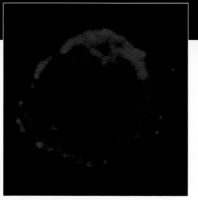

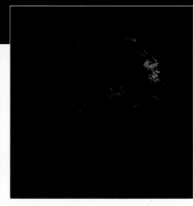

CHANDRA X-ray (high energy)    CHANDRA X-ray (low energy)    HUBBLE OPTICAL    SPITZER INFRARED

This was not only the last supernova to be seen in the Milky Way Galaxy, but it was also relatively close, being only 20,000 light-years from Earth. By taking images of the supernova remnant at a range of wavelengths, we can learn much about the surface and composition of the giant star that blasted into space, as well as the clouds of cooler gas that the exploded material collided with. This reveals details of materials with a whole range of temperatures and compositions.

## ELEMENTS IN THE STARS

We used to think that the early Universe mainly consisted of hydrogen and helium. We thought that metallic elements such as iron, calcium, magnesium, silicon, and neon were produced later in time, in supernova processes in massive evolved stars. But the Hubble has found that quasars in the early Universe have even greater iron concentrations than our Sun. Many of these metals have spectral features in the X-ray and infrared regions of the spectrum, regions that can be probed by the Chandra and Spitzer space telescopes. Data from these two instruments have confirmed the Hubble findings.

## GALACTIC ARMS

Combining Hubble and Spitzer data is extremely useful when investigating processes taking place in galactic arms. A perfect example is the Orion Nebula. In its heart is the Trapezium (see page 24), a cluster of four massive stars, only about 1,600 light-years away and each star 100,000 times more luminous than the Sun. In the cluster's neighborhood lie hundreds of less massive stars and also swirling clouds of gas and dust. Stellar winds from newborn stars and ultraviolet radiation from the massive stars interact with the hydrogen, sulphur, and carbon-rich molecules in the surrounding regions. Spectral lines seen by the two telescopes enable the distribution of these elements and molecules to be mapped.

## SOMBRERO GALAXY

Another example of telescope symbiosis is in the observation of the Sombrero Galaxy, Messier 104. This nearby spiral galaxy is only 28 million light-years away and at 50,000 light-years across it is about half the size of our Milky Way Galaxy.

When imaged by Hubble the galaxy looks like a Mexican sombrero hat seen edge on; its brilliant central bulge of stars is the hat's crown, and the galaxy's huge dust ring, the hat's rim.

Hubble can only see the nearside of the ring of dust; it appears in silhouette. By contrast, Spitzer's infrared view sees the ring extended all around the galaxy, and it reveals an otherwise hidden disk of stars inside the dust ring. The Spitzer view also shows clumpy areas in the far edges of the ring where young stars are forming. A further surprise is that the Spitzer view clearly shows the galactic disk is warped. At some time in the near past the Sombrero Galaxy has had a close encounter with another galaxy and has become gravitationally deformed.

The X-ray eyes of the Chandra telescope reveal hot gas as well as point sources that are a mixture of objects within the Sombrero. Further point sources are quasars in the far distance. The Sombrero's X-ray glow extends beyond the visible limit of the galaxy. This is possibly the result of a wind driven in the main by supernovae that have exploded within the galaxy.

ABOVE
**Sombrero by Hubble**
The Sombrero Galaxy is the largest galaxy in the nearby Virgo Cluster. This Hubble view shows a prominent rim of obscuring dust on the near side of the galaxy blotting out the central bulge of stars.

RIGHT
**Combined view**
When Hubble and Spitzer data is combined a complete ring of dust and stars are seen circling the whole galactic nucleus. This ring is slightly warped indicating that the Sombrero has had a recent close encounter with another galaxy.

RIGHT
**Three-way vision**
This image of the Sombrero is a composite of Hubble, Spitzer, and Chandra data. In addition to the details seen in the previous two pictures, we can now see an extended enveloping halo of glowing X-ray plasma. This is produced by a wind of material from galactic supernovae.

# PROBING THE PLANETS

For most of the Space Age, scientists have launched astronomy satellites to study the distant stars. To study the planets and other members of the solar system, they launch different kinds of spacecraft, commonly called space probes. Unlike satellites, probes are designed to escape completely from Earth's powerful gravity. To do so, they must be launched at what is called the escape velocity, which on Earth is a speed in excess of some 25,000 miles per hour (40,000 km/h).

Although it missed its target planet, Venus, by thousands of miles, the probe Mariner 2 made the first deep-space discovery in 1962. It revealed that the planet had a heavy carbon dioxide atmosphere. Three years later, Mariner 4 returned the first close-up pictures of a planet, showing the cratered landscape of Mars.

As the years passed, the cameras of these tiny robot explorers revealed the mysteries of one planet after another—Venus, Jupiter, Saturn, Uranus, and Neptune. To date, the dwarf planet Pluto has not been visited by any space probes, though one is now on its way. In January 2006, New Horizons started its journey to Pluto. On arrival in 2015, it will start five months of study before traveling on into the Kuiper Belt.

The planets are not the only targets of space probes. Giotto and other spacecraft carefully encountered Halley's Comet in 1986. Galileo returned the first close-up images of an asteroid (Gaspra) in 1991. And in 2000, NEAR-Shoemaker touched down on Eros (see page 128), making the first landing on an asteroid.

LEFT
**Venus Express**
An artist's impression of the European space probe Venus Express in orbit around Venus. It arrived in April 2006 and is scheduled to work until December 2014.

BELOW
**Victoria Crater**
The panoramic camera of NASA's Opportunity rover images the rim of this 2,625-foot (800-m) diameter crater that has been gouged out of the Martian desert by an impacting asteroid. Opportunity spent almost two years exploring in and around the crater, from September 2006.

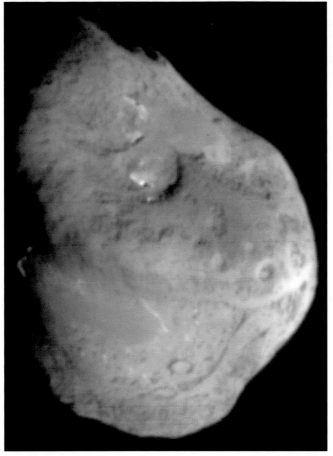

RIGHT
**Cometary encounter**
The nucleus of Comet Tempel I is seen by the Deep Impact spacecraft as it approaches. Five minutes after this image was recorded, Deep Impact's impacting probe smashed into the comet's surface.

FAR RIGHT
**Europa**
The Galileo space probe made 11 close flybys of Jupiter's moon Europa. Images returned show an icy surface crossed by dark lines, and cracks and ridges thousands of miles long.

BELOW
**Jupiter and Io**
Jupiter's moon Io is seen above the planet's colorful surface. The image was taken by Cassini on the dawn of the new millennium, January 1st 2001, as it flew on to encounter Saturn.

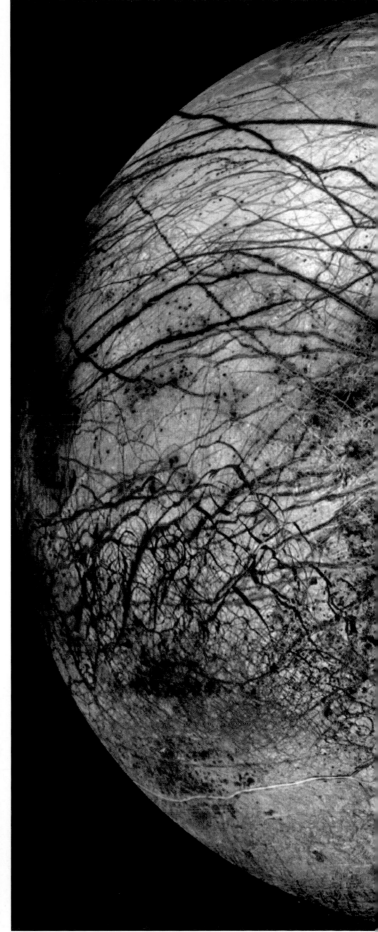

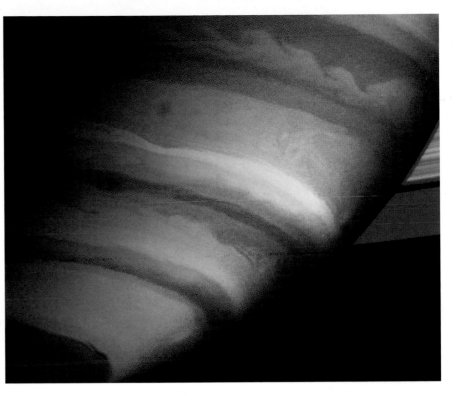

LEFT
**Lift-off!**
An Atlas V rocket lifts off from its launch pad at Kennedy Space Center in Florida on January 19th 2006. Onboard is New Horizons destined for the dwarf planet Pluto and the Kuiper Belt.

RIGHT
**Dragon Storm**
A false-color image of Saturn made from Cassini images taken in infrared light show this planet is far from serene. The reddish feature above center is Dragon Storm, a giant thunderstorm.

LEFT
**Cassini at Saturn**
Saturn's rings cast a shadow on to the planet in this March 2007 view taken by the Cassini spacecraft. One of the planet's moons, Dione, hangs in the distance beyond Saturn.

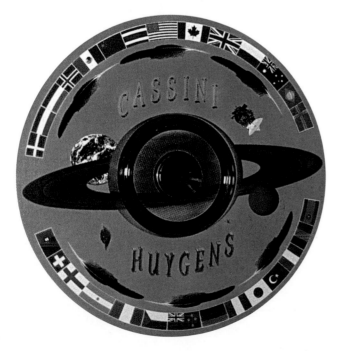

LEFT
**Message from Earth**
The Cassini spacecraft carries a DVD with 616,400 signatures of people from all around Earth. The disk is decorated with flags from 28 nations. Saturn, Earth, Titan, Cassini, and Huygens are also represented.

## STAR VOYAGER

One probe stands out for its outstanding achievements. It is Voyager 2, the first of an identical pair launched in 1977. The prime targets of both were Jupiter and Saturn, and both probes sent back incredible images of these two planets, their moons, and rings. While Voyager 1 moved off into interplanetary space, Voyager 2 continued to explore other worlds.

The Voyager mission team had decided to make a "grand tour" of the giant planets, taking advantage of a planetary configuration that would not occur again for some 176 years. After picking up gravitational energy as it passed Jupiter (1979) and then Saturn (1981), Voyager 2 went on to encounter Uranus (1986) and Neptune (1989).

## BUILDING ON SUCCESS

Before the Voyagers had completed their work, plans were in hand for a new mission to Jupiter; one that would orbit the planet rather than just fly by. From late 1995 until 2003, Galileo made a thorough study of the Jovian system; orbiting the planet and flying close to its four major moons. In 2004, 23 years after Voyager 2 had left, Cassini released a smaller probe, Huygens, which descended to the surface of Saturn's moon Titan. Cassini then started its in-depth study of Saturn and its moons, which will continue until 2017.

Meanwhile, Mars Express, Europe's first venture to Mars continues to orbit the red planet while the rover Opportunity explores its surface. Messenger orbits the planet Mercury, New Horizons continues its journey to Pluto, and the Rosetta spacecraft prepares to meet up with Comet Churyumov-Gerasimenko and release the lander, Philae, onto its surface in 2014.

# THE HUBBLE SPACE TELESCOPE

Among optical telescopes, the Hubble Space Telescope reigns supreme. It can see farther into the Universe than any others, and with greater clarity, from its vantage point hundreds of miles above Earth's obscuring atmosphere. It is named for the U.S. astronomer Edwin Hubble, who pioneered detailed study of the galaxies during the first quarter of the twentieth century using the famous 100-inch (2.5-m) Hooker telescope at Mount Wilson Observatory. The HST's mirror is comparable in size, with a diameter of 95 inches (2.4 m). But it is a much more effective light-gatherer, not only because it operates in space, but also due to its supersensitive CCDs and "state of the art" electronics.

Work began on the HST in the late 1970s, but it was not launched until 1990. "First light" revealed image-blurring due to a flawed primary mirror, and the whole project seemed doomed. But a daring recovery-and-repair mission three years later sharpened the HST's vision and imagery to design specifications. Updated systems and instruments installed on subsequent servicing missions have made the HST an astronomical observatory without equal in the study of the Universe.

LEFT
**Edwin Hubble**
The foremost U.S. astronomer of modern times, whose work led to an appreciation of the true scale of the Universe.

He might have been a world-class heavyweight boxer; he might have been a brilliant lawyer. But instead Edwin Powell Hubble (1889–1953) chose to become an astronomer.

Born in 1889 in Marshfield, Missouri, Hubble later moved with his family to Chicago, where he gained a degree in mathematics and astronomy at the University. By 1910, he was studying at Oxford University, gaining a Bachelor's degree in Jurisprudence two years later. In 1913, back in the U.S., he opened a law practice in Kentucky.

## INVESTIGATING FAINT NEBULAE

However, the lure of the heavens proved too strong for Hubble, and he returned to Chicago to become a graduate student at Yerkes Observatory. In 1917, he gained a doctorate with his thesis "Photographic Investigations of Faint Nebulae," in which he suggested that spiral nebulae might lie outside our own star system.

After a stint with the American Expeditionary Force in France during World War I, Hubble accepted an invitation by George Ellery Hale to work at Mount Wilson Observatory. It was there, in 1919, that he began observing with the newly completed 100-inch (2.5-m) Hooker Telescope, then the biggest in the world. The Hooker still wasn't powerful enough to resolve individual stars in spiral nebulae, although it could detect pinpricks of light in them that some thought might be novas.

In 1923, Hubble spotted one in a photograph of the Andromeda Nebula (now known as the Andromeda Galaxy). He then checked previous photographic plates and found points of light in the same position but with different brightnesses. He plotted the light curve of the source and found that it had the typical characteristics of a Cepheid variable. Applying Henrietta Leavitt's period-luminosity law, Hubble deduced that the Cepheid was an astonishing 900,000 light-years away.

This meant that the Andromeda Nebula was way beyond the confines of our own Milky Way Galaxy. It had to be a separate star system—a separate galaxy.

At the time the Milky Way was estimated to be 300,000 light-years across—three times its actual size. Hubble's value of the distance to the Andromeda Nebula on the other hand was a gross underestimate; the true distance is about 2.9 million light-years.

## HUBBLE'S TUNING FORK

Hubble eventually prepared a paper announcing this momentous discovery which Henry Norris Russell presented on New Year's Day 1925. Shortly after, Hubble introduced a system for classifying the galaxies, or "extragalactic nebulae," as he preferred to call them. There were irregulars, with no particular shape, and regulars, with a characteristic shape. He devised the "tuning fork" system, still used, which classifies the regulars by their shape into ellipticals, spirals, and barred spirals (see page 74).

Hubble also continued, by looking for Cepheids, to determine the distance to other galaxies. Meanwhile, at Lowell Observatory in Arizona, astronomer Vesto Slipher was estimating the speeds at which galaxies were moving by examining the spectrum of their light. He found that almost all had a red shift in spectral lines, indicating that they were rushing away from us, and at unheard-of speeds.

## AN EXPANDING UNIVERSE
By correlating his distance values with Slipher's speeds of recession, Hubble made an astonishing discovery. The farther away a galaxy was, the faster it receded. Hubble announced this startling fact in 1929, and it became known as Hubble's Law. It led to the fundamental concept of modern cosmology—that the Universe is expanding.

Just as happened in World War I, World War II interrupted Hubble's work in astronomy. He became chief of ballistics and director of the Supersonic Wind Tunnel Laboratory in Maryland. After the war, Hubble became involved with the 200-inch (5-m) Hale Telescope project at Mount Palomar Observatory. When the telescope became operational in 1949, it was fitting that Hubble should begin observations. With this powerful instrument, he continued to study the Andromeda Nebula and other spiral galaxies.

By this time, Hubble had only a few years to live. He had developed a heart condition and in 1953 died from a cerebral thrombosis. It is altogether appropriate that the Hubble Space Telescope, which is every day expanding our knowledge of the Universe, is named for the man who proved that the Universe itself is expanding.

ABOVE
**Hubble sunrise**
Sunlight turns the HST into liquid gold as the Sun peeps over the limb of the Earth, hundreds of miles away. It was one of the 16 sunrises per day the STS-109 mission astronauts experienced on their servicing mission in March 2002.

# GETTING INTO ORBIT

operational, and a committee of the National Academy of Sciences had published a report on "Scientific Uses of the Large Space Telescope." Funding for a proposed 120-inch (3-m) instrument was elusive, and so plans were drawn up for a cheaper 95-inch (2.4-m) version.

Congress eventually approved funding for the smaller instrument, and in 1977 NASA began work on the project. By then, NASA had acquired a partner, the European Space Agency (ESA). Under their agreement, ESA would provide 15 percent of the hardware, in return for 15 percent of observing time. The launch was scheduled for 1983.

## LIFT-OFF

By 1979, the detailed design and specifications of the telescope had been finalized, and construction began. With the involvement of more than 20 major contractors constructing the telescope for NASA, a number of universities and ESA developing the instruments and producing the solar rays, the building was far from straightforward. It also proved much more expensive than anticipated.

In the event, the telescope was nowhere near ready for launch in 1983, when it was officially named the Edwin P. Hubble Space Telescope. A possible 1986 launch was delayed by the grounding of the shuttle fleet following the Challenger disaster. Even when the shuttle fleet became operational again in 1988, the HST had to wait in line. Not until April 24th 1990, did the telescope ascend into the heavens. The Kennedy Space Center's launch commentator announced: "Lift-off of the Space Shuttle Discovery with the Hubble Space Telescope—our window on the Universe."

## HOW THE HST WORKS

The HST orbits the Earth about once every 95 minutes, in a nearly circular path about 350 miles (560 km) high. This height is well within reach of the shuttles that serviced it. However, this is a relatively low altitude for a satellite, and its orbit gradually decays because of faint traces of atmospheric gases still present. So, after a service mission, visiting shuttles redeployed it in a slightly higher orbit.

The HST is a big satellite, over 43 feet (13 m) long, with a diameter of 14 feet (4.3 m). It weighs about 12.5 tons (11.5 tonnes). Designed for regular servicing while in orbit, it is outfitted with hand-holds and also grapple fixtures so that it can be gripped by the shuttle orbiter's RMS (robot) arm. Its instruments are modular, so that each can be repaired or replaced without affecting the others.

Two large solar arrays, made up of thousands of solar cells, provide the HST with electricity to keep its batteries charged. The latest panels can supply more than five kilowatts. Unlike the original panels, which were flexible and unfurled, the present ones are rigid. They are also smaller, measuring about 23 feet (7.1 m) long and 8.5 feet (2.6 m) wide.

ABOVE
**Checkout**
The nearly completed HST ready for final checks. In the event, exhaustive testing of the telescope's components and systems failed to reveal a near fatal flaw in the optics.

In 1923, the same year that Edwin Hubble proved that spiral nebulae were external galaxies, German rocket scientist Hermann Oberth wrote a groundbreaking book about space travel called (in translation), *The Rocket into Interplanetary Space*. In it he discussed the possibility of a space telescope that could be attached to an orbiting space station, or even mounted on a small asteroid for stability.

Oberth's ideas were far ahead of his time. Not until 1946 did the concept of a space telescope resurface, in a paper written by the U.S. astronomer Lyman Spitzer. It was entitled "Astronomical Advantages of an Extra-Terrestrial Observatory." Later, the Soviet craft Sputnik 1 went into orbit, and space flight became a reality.

## THE LARGE SPACE TELESCOPE
In the 1960s, Spitzer began to lobby NASA and the U.S. Congress to provide funding for a large orbiting telescope. By the decade's end, NASA's first orbiting telescope (OAO-2) was

**The HST in orbit**
The main body tube houses a fairly conventional reflecting telescope, which feeds focused light into instruments in the aft assembly. Data gathered by the telescope is transmitted back to the ground by radio from a pair of antennas. Twin panels (arrays) of solar cells provide the necessary electrical power.

**Hands-on**
STS-82 astronauts Steven Smith (left) and Mark Lee gather their tools in preparation for another day's work on a routine servicing mission. The HST was the first major satellite designed specifically for in-orbit servicing.

# HST IMAGES

The stunning beauty of the Hubble images is clear to all, but the science behind them is much more prosaic and relies on considerable skill and computer power.

The images, such as those seen throughout this book, start off as monochromes. First, a view is recorded on, and then downloaded from a Charged Coupled Device (CCD)–a silicon chip–in the focal plane of a Hubble camera. The chip is similar to any found in a personal digital camera. Hubble's cameras are fitted with a series of colored filters, and pictures are taken in each of the colors. Once transmitted to Earth, these pictures are added together, using a computer, to produce the final colored images.

Many of the images are produced by simply combining three exposures, taken in red, green, and blue light. Once mixed together they give the human-eye view. The process is very similar to that used by television screens, computer monitors, and video and still cameras. The intensities of the single colour pictures can be chosen so that the final image is in exactly the same colours that we would see. It is as if we were personally out there, looking through Hubble.

Alternatively, pictures can be enhanced to produce a final image that concentrates, for example, on the infrared, which is beyond the visual range of the spectrum. Certain specific wavelengths can also be enhanced to bring out subtle details. This is possible because the HST not only detects all the visible wavelengths, but can also detect the shorter ultraviolet wavelengths, and the longer infrared ones.

Images taken by the Wide Field and Planetary Camera 2 (WF/PC2) have a stepped shape. This camera takes pictures using four component cameras; one of which records a magnified view. Once the magnified view is scaled to fit the others the image is step-shaped.

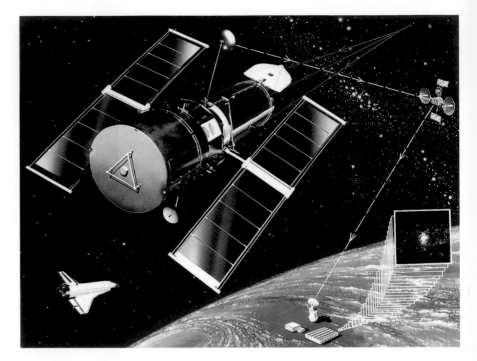

Those images that cover a large-field of view are achieved by taking a roster of pictures and then "stitching" them together, rather like assembling a jigsaw puzzle, but with square pieces. This is because the field of view of the HST's cameras is rather small.

Imagine using Hubble's Advanced Camera for Surveys. Its Wide Field Camera has a detector that consists of two 2048 x 4096 CCDs butted together, giving 16 megapixels in all. The field of view is 202 x 202 seconds of arc which is astronomically rather small. Compare, for example, the Sun and Full Moon which are each about 1800 seconds of arc across. So if Hubble was used to produce an image of the Full Moon it would have to take about 70 individual pictures and stitch them together.

ABOVE
**Data flow**
Hubble's recorded data is relayed via the Tracking and Data Relay Satellites system to a Ground Station at White Sands, New Mexico. It is then forwarded to a control center and finally to the Space Telescope Science Institute, Baltimore, for processing.

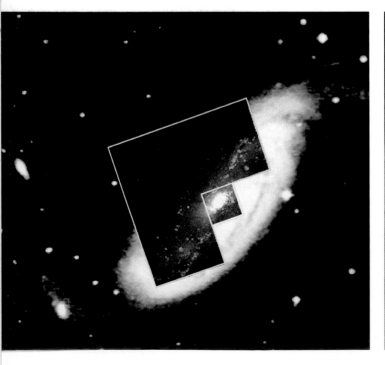

LEFT
**Stepped view**
These views are of the barred spiral galaxy NGC 1808; a starburst galaxy where vigorous star formation is taking place. At far left, a combined step-shaped WF/PC2 image is superimposed over a ground-based telescope picture. The smaller square shows the planetary camera view shrunk to the same scale as the wide-field view. The more detailed planetary camera view is also shown at near left.

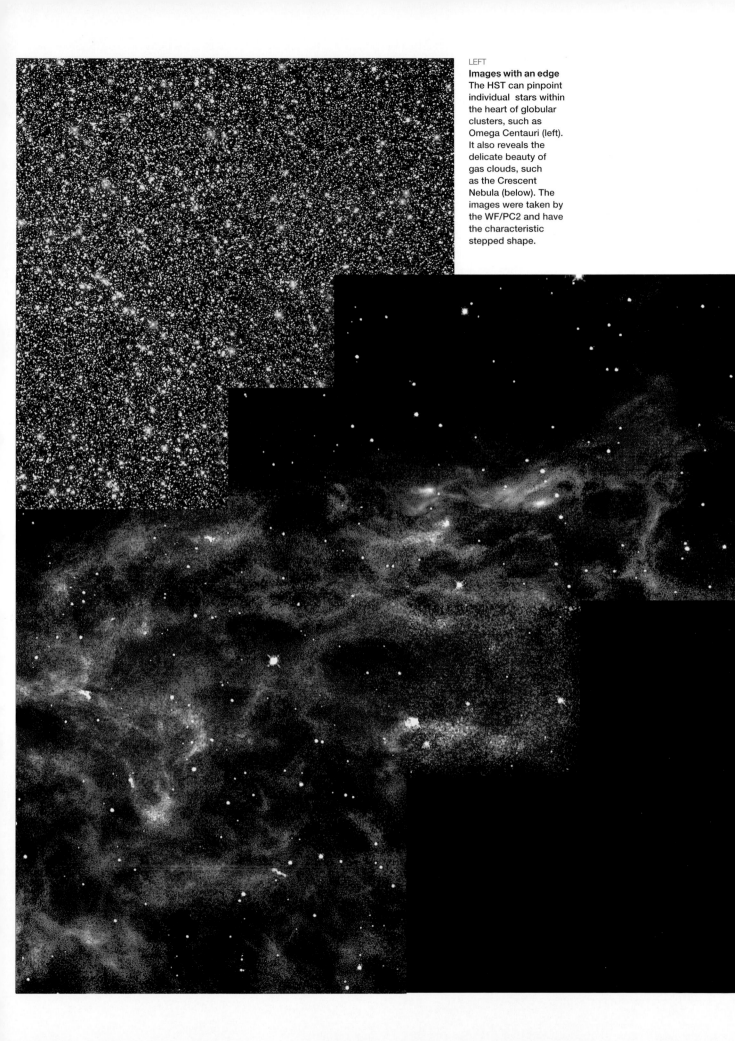

**Images with an edge**
The HST can pinpoint
individual stars within
the heart of globular
clusters, such as
Omega Centauri (left).
It also reveals the
delicate beauty of
gas clouds, such
as the Crescent
Nebula (below). The
images were taken by
the WF/PC2 and have
the characteristic
stepped shape.

LEFT
**Simulated color**
The color pictures the HST takes—say, of Mars—are not snapped in color, as we snap pictures on color film. They are snapped in a narrow band of wavelengths, selected by passing incoming light through a color filter. Reproduced in this form, they present a monochrome, or black-and-white image (left). A true-color image (right) can be simulated by combining several images taken through different color filters and assigning them appropriate colors.

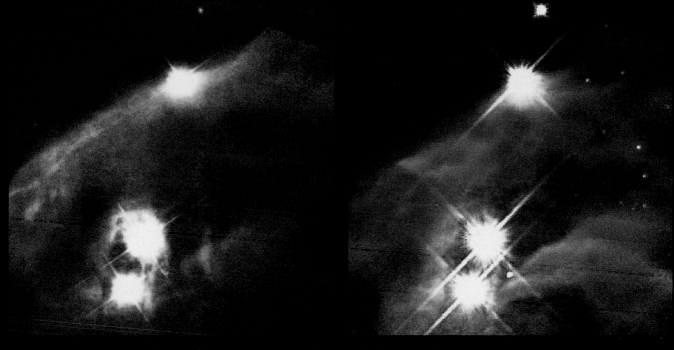

**Colorful conclusion**
Delicate filaments of
glowing gas in
Cassiopeia mark the
remains of a star that
blew itself to bits in a
supernova explosion
that astronomers
witnessed in 1572.
The HST team has
combined and colored
images taken through
the filters to highlight
the presence of
different chemical
elements. For
example, deep blue
shows regions rich in
oxygen; red, regions
rich in sulfur.

**Seeing the invisible**
These two different
views of the same
region of the Cone
Nebula show how
different wavelengths
tease out additional
details. At left is a
visible-light image
from the Advanced
Camera for Surveys;
at right, an infrared
image from NICMOS
(see page 198). Note
that infrared light has
penetrated the dusty
haze to reveal a clutch
of hot stars.

**Remote cepheids**
This remote galaxy is
NGC 4603, just one in
a cluster of galaxies
in Centaurus. The
HST has been able to
pinpoint Cepheid
variables in the
galaxy, establishing
its distance at 108
million light-years.

# A FLAWED DEBUT

Seven years after its original planned launch date, the HST arrived at the Kennedy Space Center in October 1989. It was installed in the payload bay of shuttle orbiter Discovery on March 29th 1990, with launch planned for April 10th. The HST launch mission was designated STS-31.

However, problems with the orbiter's APUs (auxiliary power units) caused the launch to be scrubbed at the T-4 minutes hold in the countdown. Two weeks later, on April 24th, the countdown went all the way, and Discovery and the HST thundered into the heavens.

Next day, with Steve Hawley at the controls, Discovery's robot arm picked up the HST and lifted it out of the payload bay. When Hawley sent commands to unfurl the twin solar panels, one unfurled but the other jammed. Inside Discovery's airlock, astronauts Kathy Sullivan and Bruce McCandless were suited up, ready to go on EVA to free the jammed solar panel manually. But eventually it freed itself.

Hawley released the robot arm's hold on the HST, which stayed close by. Two days later, it took two attempts to open the aperture door of the HST. It was at this point that Story Musgrave, the CapCom (capsule communicator) at Mission Control, radioed: "Discovery, Hubble is open for business."

## FIRST LIGHT

As it happened, the HST wasn't quite "open." The first few weeks after launch were used for equipment, systems, and engineering checks. Not until May 20th 1990, did the HST experience "first light"—this is the time when a telescope becomes fully operational and returns the first image.

This long-anticipated event unfolded before the eyes of journalists from around the world at the Space Telescope Science Institute, who cheered as an image flashed onto the screen before them. The portrait of the open star cluster NGC 3532 was actually a somewhat boring image, but it was greeted with elation by the press and with more than a little relief by mission scientists. This was the first evidence that the HST actually worked!

Behind the scenes, however, HST scientists were worried. They realized that the first-light image was blurred. And no matter what they did, they couldn't make it sharp. The HST wasn't focusing properly! Word soon leaked to the once adulatory press, who were quick to pour scorn: "Pix nixed as Hubble sees double" screamed one memorable headline.

The HST team soon realized that their 1.5-billion U.S. $ brainchild had flawed optics. The primary light-gathering mirror was suffering from a classic curved-mirror defect known as spherical aberration. The mirror had an incorrect curvature, which caused light rays from different parts of the mirror to be brought to a focus at different points.

It turned out that the primary mirror deviated from the correct curvature by a miniscule amount—just two microns, or about one-fiftieth of the width of a human hair. Still, this was enough to blur the images. The defect had apparently been caused in manufacturing and had not been detected during subsequent tests.

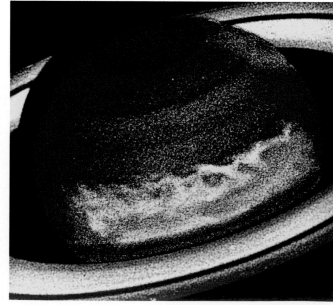

# EMERGENCY SERVICE

**LEFT**
**Capture**
It is December 4th 1993. Shuttle orbiter Endeavour maneuvers closer and closer to the ailing HST, seen here against a backdrop of the Indian Ocean off Australia's west coast. Soon the HST will be safely berthed in Endeavour's payload bay.

**LEFT**
**New for old**
On their EVA of December 7th Jeff Hoffman and Story Musgrave install the new WF/PC-2. Here Hoffman is seen with the old instrument, which he will stow in the payload bay.

So here was a telescope hundreds of miles away in space that was returning blurred pictures due to faulty optics. While NASA and the Space Telescope Science Institute debated what to do, computer experts were devising a way to manipulate the incoming data in order to improve the quality of the images. This computerized image-processing technique, called deconvolution, involved determining where light was missing from the image and restoring it.

Deconvolution and longer exposure times certainly helped to produce acceptable images, but they were still only on a par with the best images taken by the most powerful observatories on Earth under good conditions. The mammoth leap in observational astronomy that scientists were expecting from the HST had not materialized.

How to fix the HST—that was the problem. Some suggested that it should be recovered, returned to Earth, refurbished here and then redeployed. But the consensus was that the HST should be repaired in orbit. After all, it been designed for in-orbit servicing.

## COSTAR TO THE RESCUE

The Hubble team soon devised a means of correcting the telescope's defects. They would use an ingenious piece of equipment with 10 small mirrors, called COSTAR (corrective optics space telescope axial replacement). COSTAR would clarify the Hubble's vision rather like eyeglasses cure defective eyesight. Its mirrors would refocus the light coming from the defective primary and feed it to the Faint Object Camera (FOC) and spectrographs. However, in order to make room for COSTAR aboard the HST, the High-Speed Photometer (HSP) would have to be removed.

The HST team would also replace the existing Wide Field and Planetary Camera (WF/PC-1) with a "clone" (WF/PC-2) that had reconfigured mirrors. These, too, would combat the effects of the mirror's aberration.

## MISSION ACCOMPLISHED

By December 1993, NASA was ready to launch a rescue and repair mission, designated STS-61. And just in time. By now, the HST really was in trouble and on the verge of becoming useless. Three of its six gyroscopes had failed, and so had two of its memory banks. In addition, its solar panels vibrated every time it passed between day and night or light and shade, which happened 16 times every 24 hours.

Just before dawn on December 2nd 1993, shuttle orbiter Endeavour blasted off on the vital first servicing mission. It rendezvoused with the HST two days later. Using the robot arm, Endeavour captured the telescope and berthed it in the payload bay. On December 5th, veteran spacewalkers Story Musgrave and Jeff Hoffman made the first of five EVAs (extravehicular activities) to make repairs. It took them nearly eight hours to replace the faulty gyroscopes and other electronic parts.

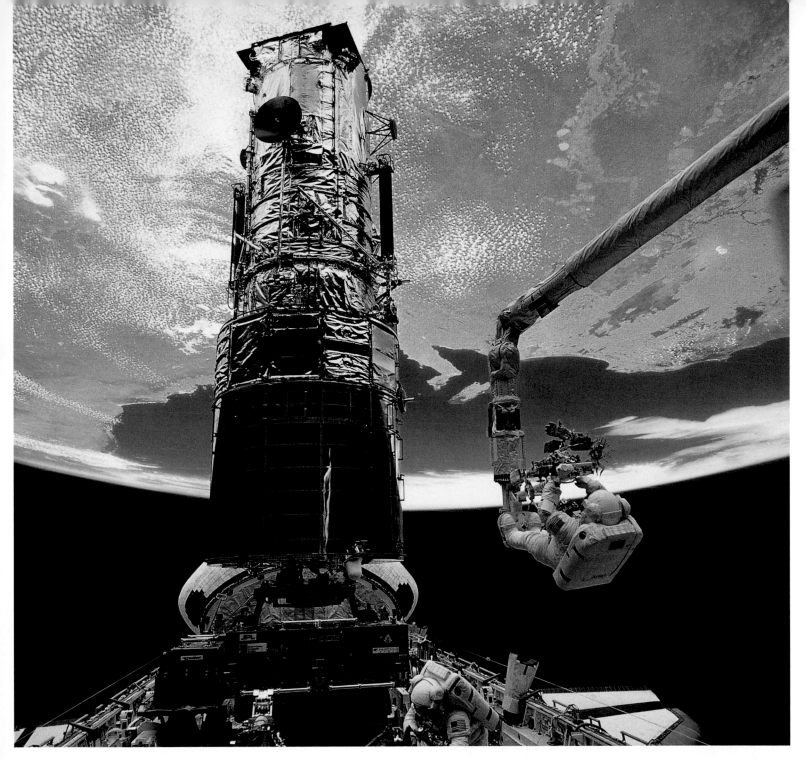

On December 6th, Kathryn Thornton and Tom Akers went on EVA to replace the two solar panels. Musgrave and Hoffman returned the next day to install the new WF/PC. On December 8th, Thornton and Akers removed the High-Speed Photometer (HSP) and fitted COSTAR into place. Musgrave and Hoffmann carried out the last of the five spacewalks later the same day, bringing the total EVA time for the mission to more than 35 hours. On December 9th, Endeavour's robot arm lifted the HST out of the payload bay and into independent orbit. Endeavour itself returned to base four days later.

## SECOND FIRST LIGHT

From NASA's point of view, the STS-61 servicing mission had gone amazingly well. But it remained to be seen whether the repair team's "eye surgery on the patient" had corrected the telescope's faulty vision. On December 18th 1993, the HST science team gathered to witness first light of the refurbished instrument. An image of a star, Melnick 34 in the 30 Doradus (Tarantula) Nebula, appeared on the screen as a bright point of light—not spread out or blurred. The repair mission had worked.

By January 13th 1994, the HST team had acquired a portfolio of spectacular images taken with the rejuvenated telescope, which they shared with the press. "The patient," an HST spokesman said, "has a new vision of incredible clarity."

## NEW VISION

Over the next four years, the HST returned spectacular images by the hundreds, proving that it had indeed opened up a new window on the Universe. In February 1997, a planned second servicing mission (STS-82) was underway. Discovery lifted off

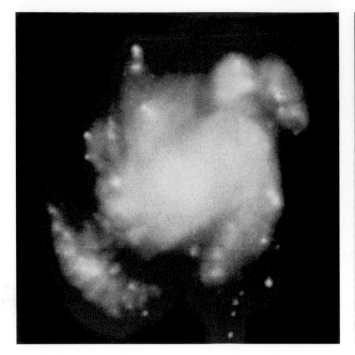
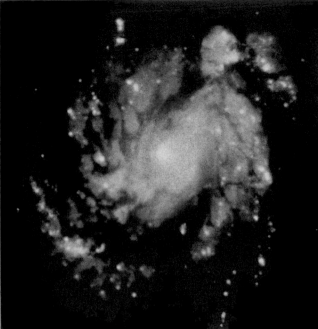

with a number of new instruments to improve the telescope's visual acuity and range.

Spacewalking astronauts, working alternately in pairs—first Mark Lee and Steve Smith and then Greg Harbaugh and Joe Tanner—carried out five EVAs to install the new equipment. On the first and most important EVA, Lee and Smith removed the two existing spectrographs and replaced them with the Space Telescope Imaging Spectrograph (STIS) and the Near-Infrared Camera and Multi-Object Spectrometer (NICMOS). The STIS would provide improved spectral resolution, while NICMOS would extend the sight of the HST into near-infrared wavelengths.

Subsequent EVAs saw the replacement of a number of instruments and equipment, including one of the Fine Guidance

Sensors (FGSs), a data recorder and a Reaction Wheel Assembly (RWA). On the final EVA, astronauts Lee and Smith repaired damaged thermal insulation on the outside of the HST, which protects the internal structure and instruments from fluctuations in temperature.

## INTO HIBERNATION

A third HST servicing mission was planned for June 2000, but one after another, its attitude-maintaining gyroscopes began to fail in 1997, 1998, and 1999. So the HST team decided to split the objectives of the mission in two, with the first launch in December 1999. On November 17th the fourth gyro failed, leaving just two gyros functioning—not enough to position the instrument. The telescope had to be shut down—it was put into "hibernation mode."

Discovery blasted off on the STS-103 service mission (designated SM3A) on December 19th 1999. On three spacewalks, astronauts replaced the faulty gyros, fitted kits to improve the telescope's batteries, and installed a new computer that was 20 times faster and had six times as much memory as the original.

## ADVANCED IMAGING

Shuttle orbiter Columbia sped into orbit on March 1st 2002, on the STS-109 service mission (designated SM3B). The crew kept a rigorous schedule in order to install all of the new and upgraded instruments and equipment.

The critical new instrument was the Advanced Camera for Surveys (ACS), which replaced the Faint Object Camera (FOC). The FOC had been "state of the art" in the 1980s, but digital imaging had improved dramatically since then.

The ACS comprises three separate instruments: the High-Resolution Channel, the Solar-Blind Channel, and the Wide-Field Channel. The High-Resolution Channel takes detailed images of the inner regions of galaxies and searches for extrasolar planets.

LEFT
**End of the arm**
Richard Linnehan is
anchored on the end
of the shuttle's robotic
arm. He is unfolding
a solar array during
the STS-109 service
mission in 2002.

RIGHT
**Whirlpool Galaxy**
The versatility of
Hubble's scientific
usefulness is
illustrated by these
two face-on views of
galaxy M51. The left
hand view (white
center) picks out the
visual radiation from
the star clusters and
star-forming regions
and was obtained
using the Advanced
Camera for Surveys
(ACS). The right hand
near-infra-red
NICMOS view (pink)
has most of the
starlight removed.
Here we see the
narrow lanes of
dense dust.

The Solar-Blind Channel, which blocks visible light to enhance ultraviolet sensitivity, studies planetary weather in our own solar system. The Wide-Field Channel helps astronomers study the nature and distribution of the galaxies on a broad scale.

The ACS was installed on the fourth of the five EVAs of the mission. Earlier, astronauts had replaced the solar arrays with a smaller but more powerful set and installed a new Reaction Wheel Assembly (RWA).

On the fifth EVA, a new neon cryocooler—a refrigeration unit—was fitted for the NICMOS. Working in the infrared, its instruments must be cooled to around minus 330 degrees Fahrenheit (–200°C). The original nitrogen-ice coolant had run out in January 1999, earlier than expected. Installation of the new cooler would bring the instrument back into operation.

The five EVAs on STS-109 broke the duration record for a single flight, totaling nearly 36 hours. Over the first four HST servicing missions, 14 astronauts had spent a total of more than 129 hours on maintenance work.

## FINAL MISSION

The fifth service mission, STS-125, which was originally scheduled for late 2005 or early 2006, very nearly did not happen. In January 2004 it, and all possible future services, were cancelled by NASA. The previous year the Columbia shuttle orbiter had been destroyed while heading home near the end of its twenty-eighth mission.

Astronaut safety constraints now imposed by the Columbia Accident Investigation Board were affecting the Hubble servicing.

There was an outcry from astronomers, especially after the release of the groundbreaking Hubble Ultra-Deep Field survey (see page 202). This image from the dawn of the Universe, just a few hundred million years after the Big Bang, underlined the fact that Hubble still had much more exciting work to do. But by October 2006 there had been very successful post-Columbia shuttle flights and so the fifth service mission was reinstated. The Hubble was to be given an overhaul and an extended life.

# NEW INSTRUMENTS

A priority of the last service mission was to add two new instruments. The first, the Cosmic Origins Spectrograph was to collect ultraviolet (UV) radiation from point sources such as stars and quasars. In the far-UV, longer than a wavelength of 115 nm (nanometer) this instrument was 30 times more sensitive than any previous space UV spectrometer. Even at around 320 nm it was twice as sensitive. It replaced the redundant COSTAR, the Corrective Optics Space Telescope Axial Replacement; the device that had been installed on the first servicing mission and corrected the blurred vision Hubble originally suffered from. More recent instrumentation had the vision correctors inbuilt so COSTAR was no longer needed.

The second new instrument was Wide Field Camera 3 (WFC3), which is sensitive to wavelengths from infrared through visual to ultraviolet. This versatile imager replaced the Wide Field and Planetary Camera 2, a camera that was also installed during the first servicing mission in 1993.

As is usual with astronomical imagers, this new instrument was fitted with a series of narrow band filters. Color pictures are achieved by taking black and white images at a range of different wavelengths and then combining these using computers. There are two CCD detectors for the ultra violet and visual. These have 2048 x 4096 pixels and work between wavelengths 200 nm and 1000 nm. There is also a single infrared detector, 1024 x 1024 pixels, sensitive to wavelengths between 800 nm and 1700 nm. This wavelength channel is similar to that which is going to be used by the James Webb Space Telescope, Hubble's successor (see page 208). Like on the Webb, this camera is being cooled by a thermoelectric cooler. The term "wide field" is rather optimistic. The width and height of the field of view is about 8.5 percent the diameter of a full Moon.

## FURTHER ADDITIONS AND REFINEMENTS

Repairs were also carried out to two instruments to bring them back to working health. The Advanced Camera for Surveys (ACS), which had taken many of Hubble's iconic images such as the Ultra Deep Field, had an electric short in 2007 and stopped working. And second, the Space Telescope Imaging Spectrograph (STIS) whose power had blown in 2004 and made it ineffective.

Additionally, certain infrastructure mechanisms were replaced. One was the Fine Guidance Sensor, which is responsible for accurate telescope pointing. A new outer blanket was added to upgrade the thermal insulation. And Hubble's six gyroscopes and its original batteries were removed and replaced by new ones too.

The replaceable systems were all easily accessible because the HST was initially designed for periodic servicing. Even though the parts varied in size from a telephone booth to a shoe box, removal and replacement took skilled astronauts only a matter of an hour or so, using special power tools.

## PREPARING FOR THE END

Another important piece of hardware was attached to Hubble. The Soft Capture Mechanism (SCM), which is the space equivalent of an automatic coupler between railway wagons, was fixed to the telescope's rear bulkhead. This will come into use when Hubble reaches the end of its useful life. An unmanned craft will then launch into a rendezvous trajectory and attach itself to the telescope using a low impact docking system. Once connected, the unmanned craft will steer Hubble into an orbit of ever decreasing size, bringing it slowly into Earth's thin upper atmosphere. Here it will eventually burn up in one glorious cosmic fireball. A large piece of "dead" space hardware, like Hubble, cannot be left in low Earth orbit—something might bump into it and be destroyed.

**BELOW**
**Repairing hardware** Shuttle astronauts Mike Massimino (left) and John Grunsfeld (right) practice the delicate repair of some of the Hubble telescope's electronic hardware during a training session at the Johnson Space Center in Texas.

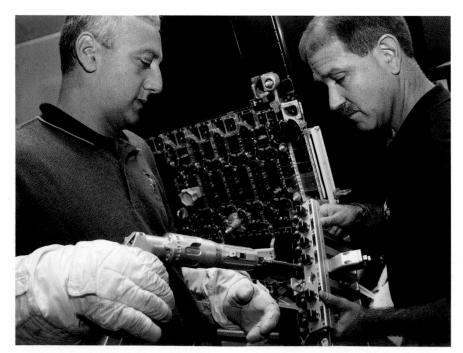

ABOVE

**Hubble hook-up**
Looking through the window of space shuttle orbiter Atlantis, May 13th 2009, we see the Hubble telescope just after it has been attached to the shuttle using the Canadian-built remote manipulator arm.

RIGHT

**Working outside**
John Grunsfeld (left) and Andrew Feustel (right) are on the first of their three spacewalks, May 14th 2009. They are preparing to remove the Wide Field and Planetary Camera 2.

# LIFT OFF TO HUBBLE

STS-125 suffered a host of delays including hurricane Hanna and tropical storm Fay. But then the telescope's Science Instrument Command and Data Handling Unit failed when tested on September 27th 2008. As this was a key piece of Hubble hardware NASA desided to postpone the launch until a new one was made and tested.

The final service mission finally blasted off from Kennedy Space Center, Florida, on May 11th 2009, just after lunch, at 2.01 pm. This was the first visit to the telescope by the shuttle orbiter Atlantis. Previous missions had been carried out by Endeavour (STS-61, December 1993), Discovery (STS-82, February 1997, and STS-103, December 1999), and Columbia (STS-109, March 2002).

Three members of the STS-125 crew had been to Hubble before. Scott "Scooter" Altman, Mike Massimino, and John Grunsfeld (twice). The other crew members were Andrew Feustek, Michael Good, Greg Johnson, and Megan McArthur. Their mission lasted thirteen days and was a complete success.

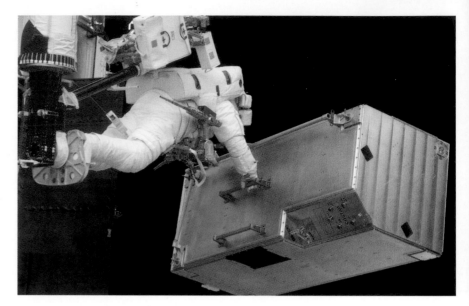

## CAPTURING HUBBLE

On arriving in orbit on day one, May 11th 2009, Atlantis' cargo bay doors were opened and the astronauts had a good look round, and then a good sleep. On day two they made a detailed inspection of the orbiter's heat shield to make sure everything was okay for re-entry to Earth. The spacesuits were checked out, and all the tools they would use when working on Hubble were assembled and checked. Day three saw the gradual approach to the HST, which was then grappled and berthed in the cargo bay.

ABOVE
**COSTAR removed**
On the third EVA (May 16th 2009) Andrew Feustel removes the Corrective Optics Space Telescope Axial Replacement unit from Hubble. This was the instrument that corrected the main mirror problems.

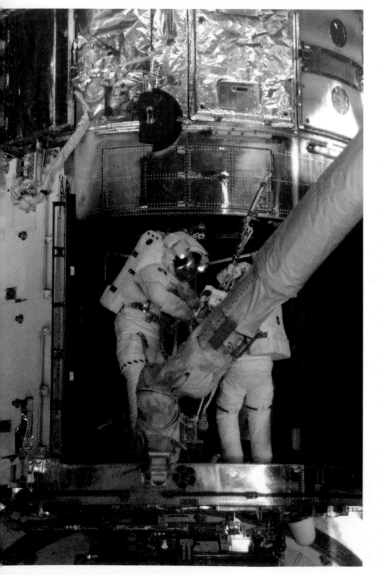

## NEW CAMERA INSTALLED

The first EVA (Extra Vehicular Activity) was made by Grunsfeld and Feustel on day four when they removed the old Wide Field and Planetary Camera 2 from the rear of the telescope. One of the attaching bolts was exceedingly stubborn. Shearing it would have been disastrous as the camera would then have to stay put. The team decided to give the bolt a real jolt with the wrench. Off it came and soon the new Wide Field Camera 3 was in place. The next job was to replace the Command and Data Handling Unit and to install the Low Impact Docking System. After seven hours and twenty minutes the two astronauts re-entered Atlantis. Overnight the new telescope camera was tested and passed with flying colors.

LEFT
**Delving inside**
Michael Good (left) and Mike Massimino are dwarfed by the telescope as they tackle the delicate problem of repairing the Space Telescope Imaging Spectrograph (STIS), an instrument that was not designed for in-orbit attention.

## SECOND SPACEWALK

On day five it was the turn of Massimino and Good. Their first task was to replace the six gyroscopes, which are mission-critical. Ever since the March 2002 servicing mission, three of the old gyroscopes had failed, another one had been switched out due to electrical problems, and the remaining two were not performing at full specification. After the gyroscopes, the batteries were replaced and another long day's work ended.

ABOVE
**The final day**
On the fifth and final day of EVA repair work, Andrew Feustel (foreground) and John Grunsfeld, prepare to replace three of the insulating thermal blankets.

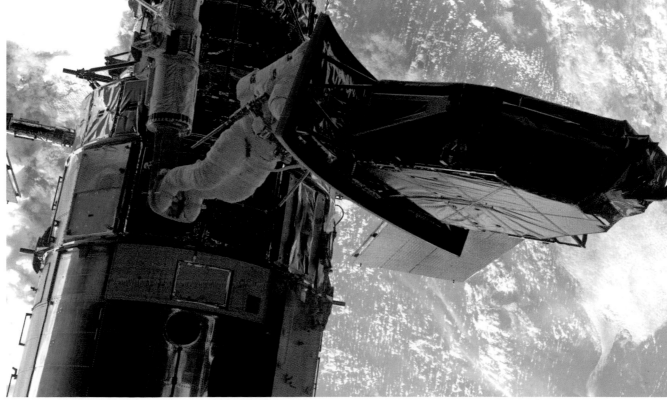

RIGHT
**New camera**
Perched perilously on the end of the shuttle's robotic arm, Andrew Feustel holds on tight to the new Wide Field Camera 3 (WFC3) on May 14th 2009. This was one of the two completely new instruments that were successfully installed during the STS-125 servicing mission.

# CHALLENGING REPAIRS

The activities on day six of the service mission were regarded as being the most challenging and uncertain. COSTAR had to be removed and the new Cosmic Origins Spectrograph attached to the telescope in its place. Grunsfeld and Feustal did this without a hitch. But then came the repair to the Advanced Camera for Surveys. This camera had failed in June 2006, and again in 2007. Also it had not been designed to be serviced and repaired in space. The two astronauts, however, sucessfully removed the access panel, replaced the four circuit boards, installed a new power supply, and closed up the system. Everything worked as planned.

The following day was equally as challenging. Massimino and Good were to repair the Space Telescope Imaging Spectrograph, another instrument that had not been designed for in-space servicing. To get at it, the cover plate had to be removed and this was held on with more than 100 screws. Each of these had to be removed, great care being taken that they did not float off into space. The job was successful but took two hours longer than expected and so the upgrading of the insulation to the outer shell of the telescope was postponed.

## FIFTH AND FINAL SPACEWALK

On day eight Grunsfeld and Feustel replaced the Fine Guidance Sensor, thus improving Hubble's ability to focus and accurately track astronomical objects. They worked so efficiently that time was left over to sort out the insulation. The last EVA was then over. All the major mission objectives had been achieved and Hubble had been given a complete upgrade. Between them the astronauts had spent a total of just under 37 hours working outside Atlantis.

## RETURN TO ORBIT

Flight day 9 (May 19th) saw the crew of Atlantis release Hubble from the cargo bay. A brief rocket burn then moved the orbiter away from the telescope. Atlantis' wings and heat shield were re-inspected for damage and all seemed fine. The next day was mainly rest and relaxation, but included conversation with President Barrack Obama and a media news conference. A further two days were filled with preparations for de-orbiting, and with worries about the weather at their Florida landing site. The weather turned out to be so unpredictable that the crew stayed in space for an extra day, finally landing on Sunday May 24th in California, at Edwards Air Force Base. After 197 orbits and a flight of around 5.2 million miles (8,368,589 km) they were home, and proud of a job well done.

## MISSION FILM

Not only did the astronauts complete all their engineering tasks, they also made a film of the procedure using an IMAX 3-D camera. The movie *IMAX: Hubble 3D* was released to IMAX cinemas in March 2010. And the crew took with them

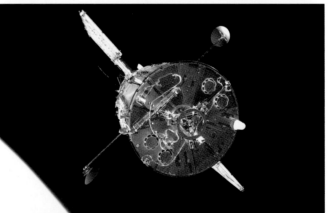

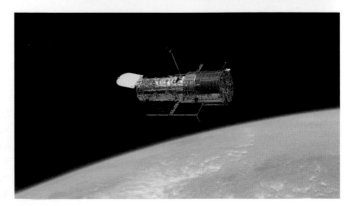

LEFT
**Leaving Hubble**
On May 19th 2009 the repaired and refurbished HST is carefully lifted out of Atlantis' cargo bay using the robotic arm (left). After being in the shuttle for a week, Hubble is released and gracefully drifts away (left middle). Here we see the base of the telescope, two of the communication antennas, and the huge wing-like solar panels. With Hubble in the distance, ready to restart its observation program (left bottom), the Atlantis shuttle team prepare to return to Earth.

two basketballs. One, associated with the Harlem Globetrotters is now in the Naismith Memorial Basketball Hall of Fame. The other was the ball actually used by Edwin Hubble, in 1909, when he played for the University of Chicago. This has gone back to his old alma mater in Illinois.

## BETTER THAN NEW

The fact that the HST could be visited in space is just one way that Hubble is special. Since its launch in 1990 its whole system has been serviced and updated, its infrastructure has been renewed, and new instruments have been added, and all on a regular basis. Routine work has included replacing

limited-life items such as gyroscopes and electronic systems. But what is more important is that old instruments have been replaced by new ones with upgraded sensitivities. Great advances have been made in electronic image recording since 1990 and Hubble has been able to keep up with them.

Hubble is now "better than new," and is good for many more years of hard astronomy. Its data is generating, on average, fourteen scientific articles a week. It could easily keep working until the JWST is up and running; working well into its third decade of operation. As the JWST only observes the infrared, as opposed to Hubble's ultra-violet-visual-infrared capability, many astronomers hope that the two telescopes might even work in tandem for a while.

ABOVE
**NGC 6217**
This image of barred spiral galaxy NGC 6217 was the first of a celestial object using the newly refurbished Advanced Camera for Surveys (ACS). Taken in June–July 2009 as part of a testing programme, it underlines the superb success of the STS-125 servicing mission.

# THE JAMES WEBB SPACE TELESCOPE

**James Webb**
James E. Webb (1906–1992) was NASA's second administrator, the man who held the reins during the early years of NASA's manned spaceflight programme, when John F. Kennedy and Lyndon B. Johnson were presidents of the United States.

The Hubble Space Telescope has been a huge success, but like any well-used machine, the time will come when it is at the end of its useful life. This problem is made worse for the HST because the Space Shuttle Program, which has sent astronauts to the telescope to carry out running repairs and change instrumentation, is being decommissioned.

In fact, no sooner than Hubble was up and running, than the next generation space telescope was being planned. Initially it was called just that, the NGST (Next Generation Space Telescope). But in 2002 it was renamed the James Webb Space Telescope, a name that was quickly abbreviated to JWST, or more easily "the Webb." The telescope is named for James E. Webb (1906–1992), a NASA administrator. He headed the Agency during the early years of manned spaceflight, being in charge of the Mercury and Gemini programmes that led to the Apollo missions to the Moon.

In astronomy, the bigger the telescope, the better; so, it is no surprise that Hubble's 95-inch (2.4-m) primary mirror is to be superseded by one 256 inches (6.5 m) across. This 2.7-times increase in size means that galaxies of similar energy output could now be detected if 2.7 times farther away. And,

as galaxies are effectively uniformly scattered throughout space, the Webb will be able to investigate 20 (2.7 x 2.7 x 2.7) times more of them than was possible by Hubble.

## WORKING IN THE INFRARED

Distant galaxies near the edge of the Universe are moving away very quickly, and their light is thus red-shifted. This means they are studied more effectively in the infrared region of the spectrum. So instead of concentrating on the visual and near ultraviolet, the Webb will be sensitive to wavelengths from 0.6 micrometers, just inside the visual red, to 28 micrometers, which is in the deep infrared.

To make sure that the radiation from distant stars and galaxies is not contaminated by infrared emissions from the telescope and its instruments, the whole JWST system is to be refrigerated to about –382 degrees Fahrenheit (–230°C). It also has to be shielded from the hot radiation emanating from the Sun, the Earth, and the Moon. This is done by placing the telescope behind a huge sunshield made of metallized foil. About the size of a tennis court, the sunshield is folded for launch but unfurls in space like a large fan.

The orbit is also important. Instead of the low-Earth circular orbit occupied by Hubble, the Webb is to be moved out along the Sun–Earth line to be near a gravitational balance point in space known as L2. Named after the French mathematician Joseph-Louis Lagrange, it is 940,000 miles (1.5 million km) away from Earth (over four times farther away than the Moon). From L2 the Sun and Earth appear close in the sky, so one sunshield will shade the telescope from the Earth, Moon, and Sun.

## INTERNATIONAL PROJECT

Like most big space projects the JWST is too expensive for just one country. The funding and running of the telescope will involve the United States' space agency NASA (National Aeronautics and Space Administration), the European Space Agency (ESA), and the Canadian Space Agency (CSA), as well as contributions from fifteen other nations.

In 2005 the total cost of the project was estimated to be about 4.5 billion U.S. $. The largest portion—3.5 billion U.S. $—was to be used to design, develop, launch, and commission the telescope and its instruments; the remainder would run the telescope for ten years. But when these ten years will start is still not clear. A combination of budget restrictions and cost overruns mean that the ealiest launch date will be 2016. The 6.2-ton (5.6-tonne) telescope will be lifted into space on board a huge European Ariane 5 rocket from the Guiana Space Center, Kourou, French Guiana, South America. After launch it will take about six months to commission the instrument, and then scientific observations will start. Just like the Hubble, the JWST will be operated by the Space Telescope Science Institute, John Hopkins University, Baltimore, Maryland.

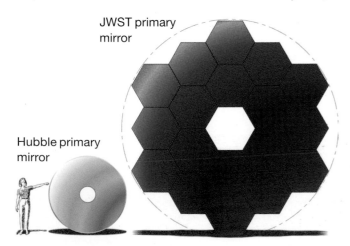

JWST primary mirror

Hubble primary mirror

**Bigger**
The segmented JWST primary mirror is 256 inches (6.5 m) across, and dwarfs the old monolithic HST ultra-low expansion 95-inch (2.4-m) glass mirror; it is 2.7 times bigger. The JWST mirror is also lightweight; it is made of gold-plated beryllium.

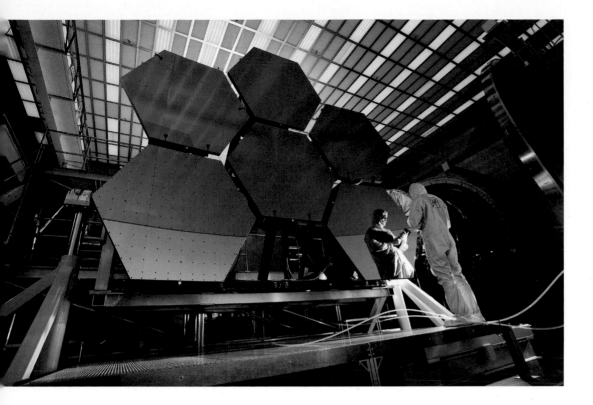

## MIRROR ON THE INFRARED UNIVERSE

The mirror design of the JWST is a major space engineering breakthrough. If the Webb's mirror was a monolithic glass mirror like Hubble's, it would be too large and too heavy to get into orbit. The JWST mirror is made up of eighteen individual 51-inch (1.3-m) diameter hexagonal segments. Each is made of beryllium and has a mass of 44 pounds (20 kg). The mass per unit area is 10 percent that of the Hubble mirror. The 18 segments fold up like the leaves of a drop-leaf table for launch and then once in space open out. Focusing of the mirror is accomplished by a set of mechanical actuators behind each segment. When operational, each mirror segment is aligned to 1/10,000th the thickness of a human hair.

ABOVE
**Mirror testing**
Six of the eighteen individual hexagonal segments of the JWST mirror are being tested at the X-ray and Cryogenic Facility at NASA's Marshall Space Flight Center in Huntsville, Alabama. They will reflect infrared radiation and will therefore be kept at a low temperature.

RIGHT
**Getting There**
The JWST mirror will be taken into space, folded up inside the nose cone of an Ariane 5 rocket. When in orbit the parts will unfold and click into place. Wave-front sensors feeding micro-motors will carefully position the mirror segments in the correct locations.

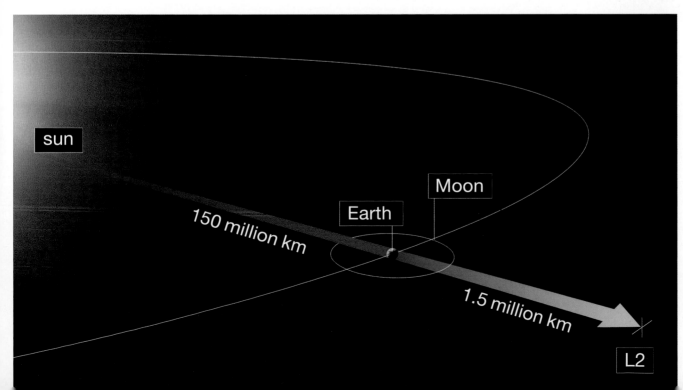

sun

150 million km

Earth

Moon

1.5 million km

L2

LEFT
**Lagrangian Point**
The JWST will be placed in space close to the Lagrangian point L2, some 930, 000 miles (1.5 million km) from Earth, directly away from the Sun on the Sun–Earth line. Here the telescope can be effectively shielded from radiation from the Sun, Earth, and Moon.

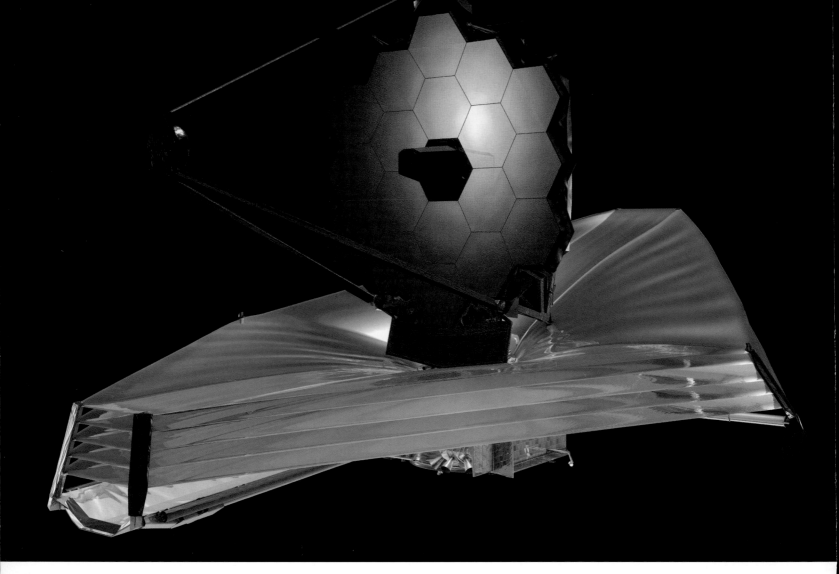

## OBSERVING GALAXIES

The main goal of the JWST is to observe the most distant galaxies in the Universe; galaxies too far away to be seen by either the largest Earth-based instruments or the Hubble. As light travels at a "mere" 186,000 miles (300,000 km) per second, the objects that we see far away appear to us as they were in the distant past. The JWST will be used to investigate very old objects, in fact the first stars and galaxies that were formed just after the Big Bang that created the Universe. The galaxies were then much closer, and collisions and mergers between galaxies more common.

## SEEING THROUGH THE DUST

There are other objectives too. The JWST is designed to collect and analyze infrared light; light that is much less affected by dust obscuration than visual light. This means the telescope can be used to probe the visually obscured center of our own Milky Way Galaxy, and the central regions of distant active galaxies. It can also be used to look into nearby gas and dust clouds, regions of space where stars are being formed, and where circumstellar disks are being produced, and planetary systems being born. These dust clouds, stellar

disks, and newly formed planets are much cooler than the surfaces of normal stars and emit most of their radiation in the infrared region of the spectrum as opposed to the visual.

## INSTRUMENTATION

The basic optics of the JWST are similar to that of the Hubble. The main mirror directs the infrared energy on to a secondary mirror which in turn reflects the infrared beam into the body of the spacecraft. Here it encounters a beam splitter that directs it to a suite of four instruments. Cryocoolers keep these instruments at temperatures of around minus 447 degrees Fahrenheit (−266°C), only a few degrees above absolute zero. The four instruments will be in place for the duration of the mission. Once in space the JWST will be too far from Earth for astronauts to replace instruments as they did for Hubble.

The Near Infrared Camera (NIRCam) is sensitive to wavelengths between 0.6 and 5 micrometers. It not only provides high angular resolution images over a reasonably large field of view but also detects the variability in the wavefront of the incoming light. This information can be used to sharpen up the images seen by the other instruments. The ten detectors in the focal plane of the NIRCam are similar to

**Webb Telescope**
The mirror of the Webb telescope sits atop the large sunshield. A pyramid of three struts holds the secondary mirror above the primary mirror, from where it reflects the captured radiation back into the telescope.

the CCDs in today's digital cameras. With a similar spectral range the Near InfraRed Spectrograph (NIRSpec) can obtain simultaneous spectra of about 100 objects. It does this by having an array of 62,000 micro-shutters in the focal plane—these being used here for the first time in space. The shutters are each 100 by 200 micrometers in size and act like minute squinting eyes. About 100 open at one time, each recording light from one of the 100 selected objects. This technique also enables the faint light of specific stars and galaxies to be spectrally analyzed even if these stars and galaxies are adjacent to brighter objects.

Being able to analyze 100 objects at once somewhat compensates for the fact that these very distant faint galaxies have to be observed for many tens to hundreds of hours before meaningful spectra are obtained. During its lifetime the telescope will obtain spectra of many thousands of the first galaxies in the Universe. The careful analysis of these spectra will give the temperature, density, radial velocity, and chemical composition of the regions being observed.

The Mid-InfraRed Instrument (MIRI) is both an imaging instrument and a spectrograph. It is sensitive to wavelengths between 5 and 27 micrometers. The camera module provides images over a wide field of view and smaller parts of that field can be examined spectroscopically at medium resolution.

The fourth instrument is the Fine Guidance Sensor Tuneable Filter Camera (FGS-TFI). The main task of this device is to lock on to guide stars and to control the fine pointing ability of the whole telescope.

## SEEING FURTHER, LOOKING BACK

The JWST will be a super-sensitive, innovative addition to the astronomer's tool bag of instrumentation. Like Hubble its results will be combined with data from other telescopes working at wavelengths varying from radio and millimeter to the visual, ultraviolet, and X-ray. The Webb will delve further back in time by looking farther out into space. It will shed more light on the "dark ages of the Universe;" the period when the first stars and galaxies were forming. And it will observe the building blocks of galaxies, and examine the birth and evolution of stars. How this new space telescope will shape up remains to be seen, but what is clear is that the HST is going to be a hard act to follow.